Routledge Revivals

The Human Experience of Space and Place

Humanistic geography is one of the major emerging themes which has recently dominated geographic writing. Anne Buttimer has been one of the leading figures in the rise of humanistic geography, and the research students she collected round her at Clark University in the 1970s constituted something of a 'school' of humanistic geographers. This school developed a significantly new style of geographical inquiry, giving special emphasis to people's experience of place, space and environment and often using philosophical and subjective methodology.

This collection of essays, first published in 1980, brings together this school and offers insight into philosophical and practical issues concerning the human experience of environments. An extensive range of topics are discussed, and the aim throughout is to weave analytical and critical thought into a more comprehensive understanding of lived experience. This book will be of interest to students of human geography.

The Human Experience of Space and Place

Edited by
Anne Buttimer
David Seamon

First published in 1980
by Croom Helm

This edition first published in 2015 by Routledge
2 Park Square, Milton Park, Abingdon, Oxon, OX14 4RN
and by Routledge
711 Third Avenue, New York, NY 10017

Routledge is an imprint of the Taylor & Francis Group, an informa business

Publisher's Note
The publisher has gone to great lengths to ensure the quality of this reprint but points out that some imperfections in the original copies may be apparent.

Disclaimer
The publisher has made every effort to trace copyright holders and welcomes correspondence from those they have been unable to contact.

A Library of Congress record exists under LC control number: 80012173

ISBN 13: 978-1-138-92462-8 (hbk)
ISBN 13: 978-1-315-68419-2 (ebk)
ISBN 13: 978-1-138-92471-0 (pbk)

The Human Experience of Space and Place

Edited by

Anne Buttimer and David Seamon

CROOM HELM LONDON

Croom Helm Ltd, 2-10 St John's Road, London SW11

British Library Cataloguing in Publication Data

The human experience of space and place.
1. Anthropo-geography
I. Buttimer, Anne   II. Seamon, David
909   GF41

ISBN 0-7099-0320-0

Reproduced from copy supplied
printed and bound in Great Britain
by Billing and Sons Limited
Guildford, London, Oxford, Worcester

CONTENTS

List of Figures

Foreword *Torsten Hägerstrand*

Acknowledgements

Introduction *Anne Buttimer* 13

Part One: Identity, Place, and Community 19

1. Social Space and the Planning of Residential Areas *Anne Buttimer* 21
2. Toward a Geography of Growing Old *Graham D. Rowles* 55
3. Identity and Place: Clinical Applications Based on Notions of Rootedness and Uprootedness *Michael A. Godkin* 73
4. The Integration of Community and Environment: Anarchist Decentralism in Rural Spain, 1936-39 *Myrna Margulies Breitbart* 86

Part Two: Horizons of Inquiry 121

5. Human Geography as Text Interpretation *Courtice Rose* 123
6. Social Space and Symbolic Interaction *Bobby M. Wilson* 135
7. Body-Subject, Time-Space Routines, and Place-Ballets *David Seamon* 148
8. Home, Reach, and the Sense of Place *Anne Buttimer* 166

Afterword: Community, Place, and Environment *David Seamon* 188

Notes on Contributors 197

Index 198

FIGURES

1.1	Location of Study Areas	28
1.2	An Operational Schema for the Analysis of Demand	30
1.3	Idealized Sketches of Social-Spatial Reference Systems	31
1.4	Idealized Activity Space Profiles	32
1.5	Standard Deviational Ellipses Describing Activity Space Profiles within the Four Estates	34
1.6	Volume of Interaction and Convenience of Destinations within the Four Estates	36
1.7	Evaluations of Site Characteristics Ranked by their Importance for the Residents	38
1.8	An Operational Model of Social Space	45
2.1	The Elderly Person's Lifeworld	62
4.1	Major Areas of Collectivization, 1936	88
4.2	Activity Patterns of Peasants Before and After Collectivization	99
4.3	Changes in the Built Environment After Collectivization	104
4.4	Generalized Scheme for Regional Collective Exchange	108
6.1	Stages of Self as Manifested in Social Space	142
8.1	Glenville, Co. Cork, Ireland	173
8.2	Glenville, Co. Cork, Ireland	173
8.3	Downtown Toronto	175
8.4	Suburbia, Chicago Region	175
8.5	McDonald's on Main Street, Worcester (or Anywhere)	176
8.6	White City (Discotheque), Route 9, Mass., USA	176
8.7	Freeway Access to Downtown Minneapolis	177
8.8	Open-pit Mine, Minnesota	177
8.9	'Old' Montreal	178
8.10	Before 'Renewal' in Glasgow Gorbals (*c.* 1965), Scotland	180
8.11	After 'Renewal' in Glasgow Gorbals (Hutchesontown, *c.* 1970)	180
8.12	'Home' in Drumchapel, Glasgow	182
8.13	'Home' in Rural Cork	182
8.14	'Housing' at 1039 Main Street, Worcester, Mass., USA	183

FOREWORD

Librarians will find it difficult to select a suitable place on their shelves for this book. Given their prejudices, they are not likely to recognize the content as geography. They will not easily find an alternative heading either. This is a fate of innovative thinking. This work is a theme with variations. Anne Buttimer and her pupils and friends direct their searchlight towards a little-explored realm: ordinary people's experience of the geography which touches the skin in daily doings and dreams. Is this theme important or just another impractical exercise? Let me give an incipient answer.

My generation believed that functional efficiency and large-scale mobility would make people rich, free and happy. We were not entirely wrong. Many good things were created, perhaps even more than we are able to put to any sensible use. But we were too enthusiastic to foresee less useful consequences: ugly, standardized landscapes; dirt in water and air; mass-media and bureaucracy; anonymous neighbors, restless children, abandoned old people. A recent reaction is that small-scale village life looks attractive to some city-dwellers. I grew up at the edge of a small and confined factory village. We gathered at the railway station to catch a glimpse of the larger world when the evening train passed by. Although I place a high price on my childhood memories of this local world with its natural splendor, I would not like to see its poverty and harsh social pecking-order established again.

At first glance this book seems to suggest the older place-bound local community as a viable solution to the social, technological and economic impass in which the urban-industrial world finds itself. My experience makes me doubt the wisdom of such a solution. To shrink systems, technologies and circulations would probably be a good thing – external circumstances seem to require such changes anyway. That is not to say, however, that all equipment, arrangements and habits of the modern world must be eliminated, and I do not think that Anne Buttimer and her group have a vision of that kind in mind. Rather, the implicit suggestion, as I read it, is that certain values almost inevitably fostered in stable, place-bound communities are indispensible ingredients for a decent human existence. There must always be a proper balance between 'place and journey,' 'home and reach.' Since we cannot return to the green valleys we remember, and do not really wish to do so, let us instead re-create the lost values by giving them visibility and inspiring people to cultivate them, each for his or her position, in the world as it now is. Once we become aware of the ailment, we may find opportunities to heal.

Torsten Hägerstrand

ACKNOWLEDGEMENTS

The people and institutions having a role in this book are many. First, the editors would like to thank the following publishers and individuals for permission to quote or reprint: Sage Publishers (Chapter 1); Westview Press (Chapter 2); George E. Deering, MD (Chapter 3); Dick Peet and Phil O'Keefe, editors of *Antipode* (Chapter 4); Croom Helm Publishers (Chapter 7); and Hans Aldskogius and Uppsala University, Sweden (Chapter 8). We also thank Hans Aldskogius for permission to use several photographs in Chapter 8.

Two university institutions have had a major role in the production of this book: first, and most obviously, Clark University, which offered the supportive intellectual milieu for the work described here; second, Lund University, Sweden, where time was provided to organize this collection amidst the business of other projects.

The editors particularly wish to thank Torsten Hägerstrand, for suggesting that this volume be assembled and providing encouragement for its completion; Christina Nordin, for criticism of various sections; and Susanne Krüger for her patience and excellent typing.

Anne Buttimer
David Seamon

Delphinen
Lund, Sweden

INTRODUCTION

Anne Buttimer

The initiative to assemble the following essays in one volume came from Professor Torsten Hägerstrand at the University of Lund in the fall of 1977. Ideas and questions which I had shared there during 1976 had aroused curiosity and concern. Issues such as environmental perception, values, subjectivity, language, stress – could these be regarded as legitimate objects for geographic study? Even if one could appreciate the humanistic or even logical grounds for such interests, how could one operationalize research on them? Often I referred to work being done by colleagues and students at Clark University and elsewhere in North America, and indeed since then there has been more exchange of ideas between Swedish and American scholars. Those who had worked directly with me did not, I felt, constitute an identifiable group: each individual had pursued his or her own line of work in conjunction with many others. In fact, we had encouraged one another to pursue topics which seemed important in their own right and none would claim the role of pioneer or spokesman for new kinds of disciplinary orthodoxy. But that is what is attractive, Professor Hägerstrand insisted: fresh beginnings and provocative theses are far more inspiring than finished products. It is in this spirit that we have responded. We present here a selection of 'excursions' – benchmarks on intellectual journeys begun at the Graduate School of Geography at Clark University and now traversing fresh territory – rather than *faits accomplis* within a unified field of expertise.

The diversity of style and orientation contained in the essays defies rigorous classification. The credit or blame for this rests largely on my shoulders and I in turn credit the milieu at Clark during those years 1970-7. I had always endeavored to make teaching an invitation to fresh discovery rather than indoctrination; I used to feel nervous and sad when students regurgitated ideas or demanded to be told what was expected of them. To work with graduate students at Clark has been for me a turbulent yet exciting journey toward discovering and reaffirming what the Socratic vocation demands in contemporary academic settings. It is with gratitude and pleasure that I join with these former students of mine in sharing some of the fruits of our time together.

Over a year of correspondence and discussion has still not yielded consensus on what an appropriate title for this collection should be. One common denominator was a shared *human experience of space and place*; the Graduate School of Geography at Clark University during those years 1970-7 was the context within which each of us drew inspiration and direction for the intellectual journeys which converged there for a while. In retrospect, given our dispositions, we could scarcely have found a more suitable context for such an endeavor. Under the Directorship of Saul B. Cohen, pluralism of thought and diversity of style could flourish. We were especially fortunate, I feel, because none of these dissertations were involved in funded Research Grants: we felt trusted to follow our own insights and to measure up to the responsibility entailed. Each person or group may describe Clark in a special way, but for me it was a setting conducive toward liberation from the tyrannies of former certainties, encouragement to reach beyond cherished taken-for-granted preconceptions and testing ground for the sincerity of one's commitment. Such learning experiences demand a high price, emotionally and spiritually, and in the rush and intensity, at times one scarcely appreciates what is happening. Only in retrospect can one begin to differentiate the wheat from the chaff.

A virtual whirlwind of associations emerge as I try to recapture the atmosphere which prevailed during those years. It was a bleak rainy afternoon in September 1970 when I arrived in Worcester, tearfully nostalgic for Glasgow and laden with data on my social space project: ill prepared for a new adventure. Graham Rowles had preceded me by one week. He too had boxes of data on a multivariate analysis of school-leavers and their perceptions of university choice, and his enthusiasm about new opportunities was radiant. David Seamon was a first-year graduate student that fall, and Bobby Wilson had already completed one year of course work.

Everything about the scene spoke of a fresh beginning: the Fiftieth Anniversary Jubilee coincided with a move to new quarters in the Academic Center. It was a lesson in proxemics to observe the social structure of the department articulate itself in the allocations of space. Environmental perception and behavior seemed to be the 'pearl of great price' at that time: bright and eager graduate students from geography and psychology pitched in to lend strength and resources to the movement. Jim Blaut and David Stea had already generated much enthusiasm over their place perception project; Ken Craik and Bob Beck were Visiting Professors in psychology for the academic year 1970-1. A faculty seminar on environmental perception and behavior included personalities as diverse as Saul Cohen, Bob Kates, Jim Blaut, David Stea,

Jeremy Anderson from geography, as well as Seymour Wapner, Bernie Kaplan, and Ken Craik from psychology.

I found myself swept up into a breathless round of discussions on those very themes which emerged in my social space project in Glasgow. Officially I was a 'post-doctoral fellow' that year, free to pursue the analysis of data and complete the report on my project. *De facto,* there was scarcely a moment free for anything except interaction with students and colleagues who seemed so eager to share ideas and questions about perception. The most absorbing of all was an attempt by a small group of us – Gerry Karaska, David Stea, Graham Rowles and I – to formulate a Research Grant Proposal which would enable us to implement a comparative study of social space in Worcester and Glasgow. In the course of many months of work we learned some unforgettable lessons about costs and benefits in translating thought into the language of research proposals. From the same experience, too, we learned much about the values of face-to-face deliberations over divergent views when the atmosphere permits frankness and mutual respect. Our proposal was not funded, but for all four of us, that group context provided supportive challenge to fresh perspectives on knowledge and life.

Other waves of intellectual interest overlapped with ours. The history of geographic thought and my interests in French *géographie humaine* provided links with the work of Martyn Bowden and Bill Koelsch, and Henry Aay helped me for two years on a course in the History and Philosophy of Geography. A major program for the training of teachers in geography, sponsored by the US Office of Education and directed by Duane Knos and Dick Ford, drew participants with a keen interest in testing some of our 'perception' models with grade- and high-school students. Bob Kates and Dick Howard shared their enthusiasm over Aquarius, a computer-based simulation of problem-solving on environmental management, and Roger Kasperson's interest in urban politics linked closely with ours.

It was not all analysis paralysis, though. Ben Wisner's vision of a journal for critical thought – an *Antipode* – was already touching many ears. This was to become, after Ben's departure for Africa, an organ for the articulation of many other views, particularly Marxist and anarchist ones. In fact, by 1972, as anti-war rhetoric waned, and the various early shoots of perception and cognition became harvested into discrete projects and dissertations, much energy became channeled toward matters of social involvement. Here a lively debate developed between 'our' existentialist vantage point and that of the revolutionary theorists, among whom David Harvey was the leading voice. On any debate between intellectual socialists and intellectual existentialists, the latter, it seemed, were inevitably the losers. My defense was usually based on logical as well as experiential evidence, but on socialism I lacked any lived experience. I think it may have been precisely this challenge which provoked me to study and live in Sweden later on. As key links between these two stances, Myrna Breitbart and Mick Godkin were especially welcome in 1972 and 1973: for both Dick Peet and for me these two persons played an enormously important role. This was my time for speculating about values for the Commission on College Geography, and in retrospect I realize that without the encouragement and constructive criticism afforded by every one of my colleagues, I doubt if I should ever have ventured into this area 'where angels fear to tread.' In the final drafts, the careful critique afforded by Martyn Bowden, Dan Amaral, Gary Overvold and Denis Wood was a lesson in scholarly cooperation and mutual support.[1]

By 1974, Graham's research on the elderly had begun, and Bobby Wilson had completed his dissertation on the experience of Black migrant families to New York.[2] It seemed that our 'urban social' phase was winding down. David Seamon had become more involved in phenomenology and had experimented with empirical observations with groups of students.[3] Myrna Breitbart was already involved in field work on anarchist communes in Spain, and Mick Godkin was spending a good deal of time studying alcoholics under the supervision of a psychiatrist.[4] In the fall of 1974, Courtice Rose arrived from Canada. Already attuned to most of the literature in urban and social geography, Curt was eager to explore the philosophical foundations of a 'new paradigm' for geography. In his capacity to read and articulate difficult nuances of analytical philosophy, Curt taught a lesson to all of us and forced us to be more rigorous in our use of language and terminology.[5] Paul Kariya joined us in 1975 with an interest in native Americans and welfare policy, and one year later Ruth Fincher, whose interests were more explicitly urban. We were indeed a motley gathering: our interests, personalities, intellectual styles, and attitudes toward social relevance were indeed varied. As I left for Sweden for the first semester of 1976 I could not have dreamed how much they were to offer help and guidance to one another.

The following essays offer a sample of the orientations which were pursued during that period 1970-7. Despite their diversity, there are some common themes discernible. There is the dialectic of security and adventure as typified in the relationship between 'home' and 'reach,' 'place,' and 'journey.' There is also the dialectic of 'manager' versus 'client,' 'movement' versus 'structure.' Throughout, also, we all groped toward a language which might permit a more sensitive relationship between 'insider' and 'outsider,' between 'supply efficiency' and 'demand appropriateness' in the organization of public services. In each of the dissertations from which these essays are drawn, either fully or in part, there is a concerted attempt to probe the experiential grounding of concepts like place, community, encounter, at-homeness, movement, and commitment.

There are limitations, of course, in the printed word of short essays: they cannot communicate much of the important learning which occurs in the course of doctoral studies. There are hazards also when one assumes the 'Cinderella' cause of human personhood and authenticity in a world where securing bread and butter often demands obliviousness to such matters. Once having switched on to an existentialist perspective, it is difficult to avoid polarizing contrasts between 'insider' and 'outsider:' this generates language and symbolism which has an anti-scientific, anti-managerial tone. Quite unwittingly, too, the humanistic intent can become vulnerable to charges of jargonizing and manipulative tokenism.

How was one to anticipate such a turn of events? In 1970 my own lenses were molded by years of philosophical immersion in existentialist and phenomenological thought and bolstered by two years of concern about planning policy and standards for housing and health in the UK. I felt a strong sense of urgency about yielding an articulate description of human experience, and the whole atmosphere at Clark at that time seemed supportive. Eight years later I wonder about the mixed blessing of 'success' in convincing so many students to pursue that course without also leading them in the direction of critical evaluation. To be a teacher is an awesome responsibility: inevitably, one's students are subjected to influences which may be intellectually exciting and inspiring but not necessarily those which will guarantee status or career advancement.

It was with a somewhat apologetic tone that I shared these worries with this group a few months ago and their response was unanimously reassuring: they too have come to recognize the connections between language, knowledge, and power within academic circles and the enduring challenge of 'vocational' versus 'system-defined' agenda for the professional geographer. To each the challenge reveals itself in different forms, for each the responsibility to face it as his or her situation allows. For all of us this volume will be a souvenir to keep the vision alive.

Notes

1. Anne Buttimer, *Values in Geography*, Commission on College Geography Resource Paper No. 24 (Association of American Geographers, Washington, DC, 1974).

2. Graham D. Rowles, 'Exploring the Geographical Experience of Older People' (PhD dissertation, Clark University, Worcester, Mass., 1976). Published as *Prisoners of Space? Exploring the Geographical Experience of Elderly People* (Westview Press, Boulder, Colorado, 1978); Bobby Wilson, 'The Influence of Church Participation on the Behavior in Space of Black Migrants within Bedford-Stuyvesant: A Social Space Analysis' (PhD dissertation, Clark University, Worcester, Mass., 1974).

3. David Seamon, 'Movement, Rest and Encounter: A Phenomenology of Everyday Environmental Experience' (PhD dissertation, Clark University, Worcester, Mass., 1977). Published as *A Geography of the Lifeworld* (Croom Helm, London; St Martin's Press, New York, 1979).

4. Myrna Margulies Breitbart, 'The Theory and Practice of Anarchist Decentralism in Spain, 1936-1939: The Integration of Community and Environment' (PhD dissertation, Clark University, Worcester, Mass., 1978); Michael A. Godkin, 'Space, Time, and Place in the Experience of Stress' (PhD dissertation, Clark University, Worcester, Mass., 1977).

5. Courtice G. Rose, 'The Concept of Reach and its Relevance to Social Geography' (PhD dissertation, Clark University, Worcester, Mass., 1977).

Part One

IDENTITY, PLACE, AND COMMUNITY

1 SOCIAL SPACE AND THE PLANNING OF RESIDENTIAL AREAS*

Anne Buttimer

> Cities are an immense laboratory of trial and error, failure and success, in city building and city design. This is the laboratory in which city planning should have been learning and forming and testing its theories. Instead the practitioners and teachers of this discipline . . . have ignored the study of success and failure in real life, have been incurious about the reasons for unexpected success, and are guided instead by principles derived from the behavior and appearance of towns, suburbs, tuberculosis sanatoria, fairs, and imaginary dream cities – from anything but cities themselves – Jane Jacobs (1961 : 6).

The livability of residential environments has become one of the most urgent challenges facing our industrial cities. Despite the volume of scientific research, experimentation, and evaluation, our understanding of the problem remains embarrassingly incomplete. Its very complexity baffles the investigator. One merely carves out slices of the problem and investigates them according to the concepts and procedures of specific disciplines.

Traditionally, residential areas have been studied within the framework of urban land-use structure (Alonso, 1964; Muth, 1969). Norms and guidelines have been developed for the 'rational' allocation of space and service functions throughout such areas (Harvey, 1970). Of late, serious efforts have been made to explore the problem from the viewpoint of the resident. Studies have attempted to explore the dynamics of spatial behavior in microenvironmental settings (Proshansky *et al.*, 1970), and several design implications have emerged from such behavioral research (Sommer, 1969; Alpaugh, 1970).

These studies also yield potential implications for planning of residential environments, but they are not yet readily translatable at that scale. Little substantial evidence is available regarding criteria on which the appropriateness of residential area design for different kinds of residents could be defined. Some studies suggest that there are

*This essay originally appeared in *Environment and Behavior, 4* (1972), 279-318. The editors wish to thank Sage Publication for permission to reprint it here.

important relationships between physical design and social behavior (Young and Willmott, 1957; Rainwater, 1966; Schorr, 1963; Yancey, 1971); others hold that little or no relationship is found between architectural design and social life (Gutman, 1966; Wilner *et al.*, 1962; Gans, 1961). Confusion abounds partly because there is still no comprehensive framework within which research on different facets of the question can be coordinated and comparative studies implemented. This multidisciplinary research effort cannot as yet claim any unifying conceptual structure, nor has it a common language for interdisciplinary effort.

Meanwhile, the planner, charged with the responsibility for designing residential environments, combs through this literature for insight into practical issues, often only to abandon it, finding common sense, traditional 'standards,' or political pressure better guides for action than 'scientific' research (Reade, 1969). Besides, the Ivory Tower ethos that has traditionally separated the planner from the academic world still constitutes a serious barrier to fruitful communication (Gans, 1968; Blair, 1969; Buttimer, 1971). Yet even when a *rapprochement* occurs, as has indeed happened on occasion in the context of residential area planning, both social scientists and planners find themselves constrained by a predominantly Cartesian view of knowledge and by the peculiarly managerial perspective on urban life which this view has fostered. Both tend to think of systems, of states of being, whether on the demand side (behavior patterns, interaction networks) or on the supply side (service networks, building design).

Livability, if this be our aim, cannot be defined adequately in terms of systems or states of *being*. For life in residential areas involves a dialogue of behavior and setting, of demand and supply; it is thus essentially a condition of *becoming*. Such a condition is seen to arise when resident communities engage in creative dialogue with their environments, molding, re-creating and eventually appropriating them as home. In this existential view, the planner can no longer be considered solely as the manipulator of supply; neither can the academician be seen merely as the investigator of resident aspiration and satisfaction. Least of all can the citizen be considered a passive pawn of external social or technological processes. This view demands that all engage themselves responsibly in the planning process itself.

For such a joint involvement in the *becoming* of residential areas, a radical new education is needed for both planner and social scientist. Each has to develop a more comprehensive understanding of urban life and the dynamics of urban systems. We need frameworks for investigation and reflection which do not segment and ossify parts of the city, as Cartesian practices have done. And we need an empathetic understanding of urban life as existential reality, as lived experience. An existential view of livability challenges the traditional rift between theoretical and applied disciplinary orientations. It calls for a unified, interdisciplinary approach to the study of environmental experience. Its essential focus on the meanings of phenomena in lived experience radically questions the assumptions and premises on which 'objective' scientific analysis is traditionally based, and openly invites subjective involvement in the reality to be investigated.

This paper addresses itself to that manifold challenge. First, it attempts to define and clarify the notion of social space as a framework for a comprehensive understanding of environmental experience. Second, it applies this idea to residential area-planning as illustrated by a preliminary investigation within selected housing estates in Glasgow, Scotland. The essay is intended to be provocative and suggestive; it does not offer rigidly tested hypotheses or guidelines for general application. Its aim is to raise rather than to resolve issues, to elicit curiosity rather than to provide conclusive answers.

Toward a Definition of Social Space

The concept of social space, as defined by Chombart de Lauwe (1956, 1952; Buttimer, 1969), offers a useful initial guide for an investigation of lived experience. As explained in his original Paris study, social space *(l'espace social)* is a framework within which subjective evaluations and motivations can be related to overtly expressed behavior and the external characteristics of the environment. Recent developments in sociology, social psychology, and other disciplines have greatly facilitated the analysis of specific dimensions of social space as defined in these terms.

In Anglo-American writings, however, semantic confusion surrounds the notion of social space. Sorokin (1928: 6) used the term to identify a person's 'relations to other men or other social phenomena chosen as "points of reference."' Social space was defined as a system of coordinates whose horizontal axis referred to group participations and whose vertical axis referred to statuses and roles within these groups. Such a 'system of social coordinates' could, in Sorokin's view, enable us to define the social position of any man. This purely sociological definition of social space differed from the psychologically oriented definitions of the term employed by other scholars, who stressed the subjective dimensions of reference systems (Park, 1924; Bogardus, 1925).

More recent definitions of the term favor the psychological orientation. One recent statement (Theodorson and Theodorson, 1969: 394), for example, holds that 'social space is determined by the individual's perception of his social world, and not by the objective description of his social relationship by any observer.' This definition implies a close connection with reference group theory, a body of literature that provides useful insights into the nature of environmental behavior (Shibutani, 1955; Hyman and Singer, 1968). These interpretations reiterate the original Durkheimian sense of the term, which defines a person's position in 'sociological space,' and specifies nothing about his situation in physical space. This was the critical link provided in the work of Sorre (1957) and further elaborated by Chombart de Lauwe (1952; 1965) in his study of Paris.

Chombart de Lauwe (1952: 190-1) identified two distinct components of social space: (1) an objective component, 'the spatial framework within which groups live; groups whose social structure and organization have been conditioned by ecological and cultural factors,' and (2) a subjective component, 'space as perceived by members of particular groups.' Recent research by Anglo-American scholars has advanced our understanding of these two components, but little attempt has been made to integrate them into any comprehensive explanatory model.

Social area analysis provides one obvious approach to a definition of objective social space. Social spaces originally denoted groupings of census tracts which displayed a degree of homogeneity in terms of given sociodemographic characteristics (Shevky and Williams, 1949; Shevky and Bell, 1955). This interpretation was later adopted by geographers for factorial ecology studies (Berry and Horton, 1970; Brown and Moore, 1971). Whether or not the 'spaces' derived from a factor analysis of census variables were considered to be 'areas' by the resident population was not considered. Pioneers of social area analysis studied the isomorphism of social participation patterns and social spaces and matched activity patterns with the spatial morphology of social characteristics (Bell, 1959; Greer, 1956), but they made little attempt to examine the isomorphism of place identification with so-called 'social spaces' (see Greer, 1969: 99-104). The analysis of social activity patterns offers more dynamic variation on this theme. Action spaces, activity spaces, behavior fields, and other concepts related to spatial movements have been examined as indices of social space (Chapin and Hightower, 1966; Cox and Golledge, 1969; Adams, 1969; Brown and Moore, 1971).

In these studies, the nature and dynamics of people's movements in space are taken as critical clues to their relationships with their environments.

Complementary perspectives on spatial experience are afforded by the literature on territoriality (Altman, 1970; Lorenz, 1966; Ardrey, 1966; Suttles, 1968), personal space (Sommer, 1969), and proxemic behavior (Hall, 1966). Processes whereby individuals and groups lay claim to space and organize and defend it in culturally prescribed ways have recently become a major focus of interest in studies of environmental behavior. Whereas research on social areas and activity spaces generally relates to 'objective social space,' the territoriality literature adds insights of 'subjective social space' (Boal, 1969; Metton, 1969). The subjective component of social space has been explored primarily in social psychology, anthropology, and ethology, within the framework of such concepts as life space (Lewin, 1951), ethnic domain (Barth, 1956), cognitive maps, (Downs, 1971; Blaut and Stea, 1971), and urban images (Lynch, 1960; Strauss, 1961). These studies, though diverse in their approach, share a common focus on perceptual and cognitive evaluations as determinants of spatial meaning (Stea and Downs, 1970; Cox and Zannaras, 1970). Such research seldom attempts, however, to link cognitive structurings of space with the actual ecological characteristics of the environment. Exceptions include Lee's (1968) empirical study of 'socio-spatial schemata,' Michelson's (1966) analysis of life-styles and value orientations, and Fried and Gleicher's (1961) work on 'satisfaction' among relocated families in Boston. Each of these studies attempts to link value orientations, mental schemata, or traditions to externally manifest behavior in particular environments.

Can any common threads of meaning be derived from these diverse bodies of literature? Is there any comprehensive framework within which they can be integrated? Are the conceptual and methodological approaches so distinct that research coordination is impossible? The literature reviewed appears to offer insight into at least five distinct levels of analysis:

(1) a social-psychological level investigating a person's position within society – that is, 'sociological space';
(2) a behavioral level investigating activity and circulation patterns – that is, 'interaction space';
(3) a symbolic level investigating images, cognitions, and mental maps;
(4) an affective level investigating patterns of identification with territory;

(5) a purely morphological level, in which population characteristics are factor-analyzed to yield homogeneous 'social areas.'

To appreciate fully the patterns yielded by any one level of analysis, they must be related to the other levels. But before any comprehensive framework can be formulated, it is important to identify some of the missing links in the chain of research endeavor.

Incorporating a Sociological Dimension

An examination of the literature on spatial behavior suggests that one critical missing link is the sociological dimension. Most of the explanatory models rest heavily on generalizations about relationships of organisms to their environments; for example, perceptual/cognitive processes (image formation, distance and space perception); dynamic-movement processes (activity spaces); instinctual/cognitive processes (territorial defense, proxemic behavior); affective processes (identification with place); and various combinations of these. The sociological dimension in these processes is rarely given explicit attention. Similarly, life-style, social stratification, status, and role are rarely treated explicitly in studies of environment behavior (see, however, Gerson, 1972).

If environmental behavior is taken as the external (spatial) expression of social reference systems (sociological spaces), it becomes possible to integrate findings from the various levels of analysis. While genetic endowment, personality attributes, territorial instincts, and so on must be recognized in any study of environmental relationships, such personal characteristics are usually influenced by the individual's life-style, group participations, and other activities involving interaction with others. The reference groups from which an individual derives his values and behavioral norms dictate certain aspirations and attitudes toward his milieu (Flachsbart, 1969; Rothblatt, 1961).

Investigation of the spatial expression of such reference systems requires an examination of spatial activity patterns generated by social interaction. A person's accessibility to social contacts, whether voluntary (friends, relatives, recreational centers) or involuntary (shops, schools, clinics), constitutes a set of congruence indices between his socially determined aspirations and his manifest behavior. The nature of a person's social relationships predisposes him to attach different significance to the routes taken, to the nodes at which interaction occurs, and to the places associated with particular events and circumstances. Discrepancies between an individual's socially dictated aspirations and his actual achievements may lead to anomalies in his spatial behavior, explicable in terms of social reference systems rather than of personality attributes, or characteristics of the environmental setting (Runciman, 1966). For each social group, a network of preferred places, interaction spaces, safe and dangerous locales, and frequented and avoided paths could be mapped. Individuals and groups feel their way through a city in activity space orbits with the nature and extent of circulation patterns generating and influencing images and establishing affective relationships with particular places, routes, and nodes.

Such sociospatial reference systems can be viewed as filters through which the physical environment is known, evaluated, and used. Geodesic space is expanded and contracted by the ties of kinship, language, and special interests. Shops, schools, and churches stand out as focuses in the mental maps of their clientele. Distances shrink or expand according to the frequency of use and the importance of destinations. 'Behavior settings' (Barker, 1968) and 'situated activity systems' (Goffman, 1961: 8) are defined in terms of the life-styles of their users. In sum, places and spaces (areas, nodes, pathways, edges) assume spatial dimensions that reflect the social significance they have for those who use them (Strauss, 1961; Lynch, 1960).

Empirical Illustration: Critique of Residential Area-Planning

The notion of social space, formulated in this way, provides a useful framework for exploring a variety of urban problems. A preliminary investigation of residents' evaluations of different housing estates in Glasgow offered an opportunity to test the idea and to develop a widely applicable methodology. The detailed research design will not be described here. Only those aspects of the study that bear primarily on operationalizing the social space concept will be discussed.

The primary aim of the study was to assess the conventional standards used in residential area design by examining residents' attitudes and evaluations of the design and service provision in selected housing estates. Planning standards recommended for residential areas concern such criteria as optimal density, accessibility to various services, design norms for house size and layout, and safety. Such standards are generally based on tradition or estimates of average demand rather than any exploration of the subjective social spaces of residents. Even when standards are comprehensive, residents of well planned estates are by no means always satisfied (Jacobs, 1961; Hole, 1959; Fried and Gleicher, 1961). This raises the question of whether the 'objective' standards approach to residential area-planning in fact stems from warrantable assumptions about social behavior.

The central question in the Glasgow study was the degree of correspondence between residents' aspirations and values (subjective social space) and standards for the design of the physical environment (objective social space). It was felt that the appropriateness of particular area designs could best be gauged by ascertaining the extent to which residents achieved the socially determined aspirations implicit in different types of spatial experience. Seen in this light, the relocation process involved more than a change of physical environments and losses and gains of services; it potentially ruptured bonds to place and to social networks. The adequacy of new environments might thus be evaluated in terms of residents' abilities to recreate satisfactory social space patterns in their new environments.

Out of 18 districts originally chosen on the basis of (a) presence or absence of planning standards, (b) location *vis-à-vis* city center, and (c)

Figure 1.1: Location of Study Areas

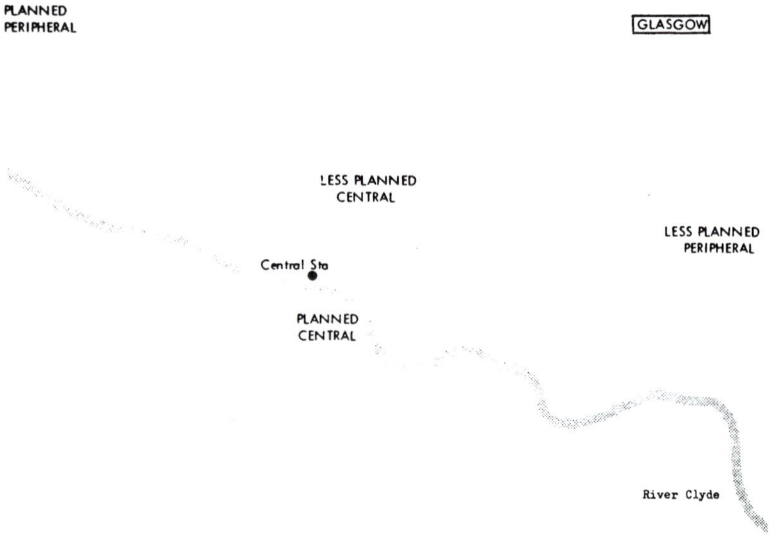

socioeconomic level as defined by ratable property values, four districts were eventually selected for the pilot study. All were in the lowest socioeconomic category; two were located near the city center, two on the periphery. One district from each location type evidenced the presence of planning standards; the other lacked them. The labels 'planned' and 'less-planned' identify these districts (see Figure 1.1).

Two distinct analytical perspectives were assumed in the study. One focused on the appropriateness of environments for people by eliciting direct assessments of actual site characteristics; the other focused on the nature of people's demands for residential environments, by exploring certain dimensions of behavior and aspirations.

The appropriateness of environmental supply was analyzed with regard to three major questions:

(1) Were residents in planned districts in general more satisfied[1] than residents of less-planned districts?
(2) How did the evaluations of external observers compare and contrast with residents' evaluations of the same characteristics?
(3) Could residents' satisfaction with their environment be inferred from the evaluations of the external observers?

The environmental assessments of residents and external observers revealed interesting points of divergence and convergence. The presence of standards did not guarantee universal satisfaction. Contrasting evaluations were largely a function of the lenses through which objects were perceived. For the external observer, notwithstanding his efforts to achieve objectivity through the standardized scales and disciplinary research models, also evaluates the environment subjectively through the variegated prism of his experience, just as the resident does. The reference systems that influence the external observer's evaluations offer a valuable avenue for further research (see Craik, 1970).

The second approach, labelled 'demand anticipation,' explored residents' evaluations of site character in terms of their underlying socio-spatial reference systems. We found that the inhabitants used, interpreted, and evaluated their residential environments through the filter of their socio-spatial reference systems, operationally defined in this study in terms of three components; (1) territoriality, (2) activity orbits, and (3) expectations about site character (Figure 1.2). Overall satisfaction with the physical characteristics of the area and with life in the area are the cumulative result of congruence in three components of spatial experience – namely, ability to identify with a home ground, accessibility to aspired social and service destinations, and a perception of the architectural environment corresponding to an image of an ideal environment.

Figure 1.2: An Operational Schema for the Analysis of Demand

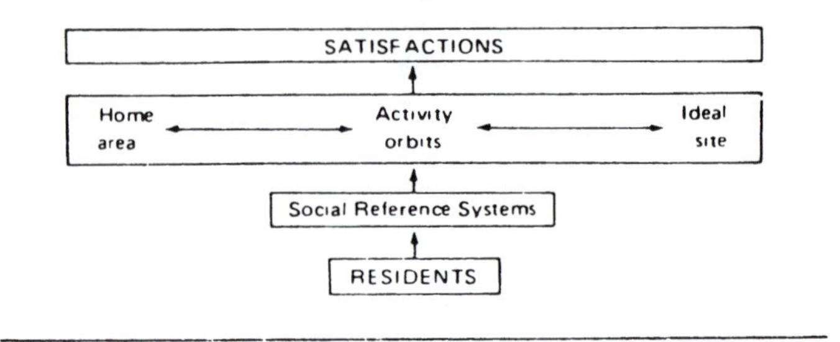

Although these three types of spatial experience can be treated separately for purposes of analysis, they cannot be considered separate entities in terms of lived experience. An image of an ideal environment theoretically subsumes aspirations about territoriality, accessibility to desired destinations, and ideal site character. Activity orbits (the spatial expression of social reference systems) contribute to and mold this image. Individuals establish affective relationships with particular urban places, routes, and nodes through the spatial activity patterns generated by interaction with their social reference systems. Their expectations about residential area character are also influenced by the norms and values transmitted within this reference system.

These relationships are illustrated for two hypothetical polar types of resident: the 'localite' and the 'urbanite' (Figure 1.3). Two distinct types of activity space were measured: (1) participation spaces defined by reference groups, shown here for relatives, friends, and occupational and special interest groups; and (2) circulation or interaction orbits, defined in terms of macro-service spaces represented by trips to schools, shops, post offices, and doctors; and micro-service spaces, which include trips to bus stops, public telephones, nursery schools, play area, pubs, youth clubs, community centers, libraries, parks, and gardens. Participation spaces are clues to a person's sociological space, while his circulation orbits are clues to his behavior field.

Three distinct layers of socially significant spaces can be defined for each individual: social participation, macro-service, and micro-service. The degree of overlap among these layers reflects the internal homogeneity or restrictiveness of an individual's social space. In spatial terms,

Figure 1.3: Idealized Sketches of Socio-spatial Reference Systems

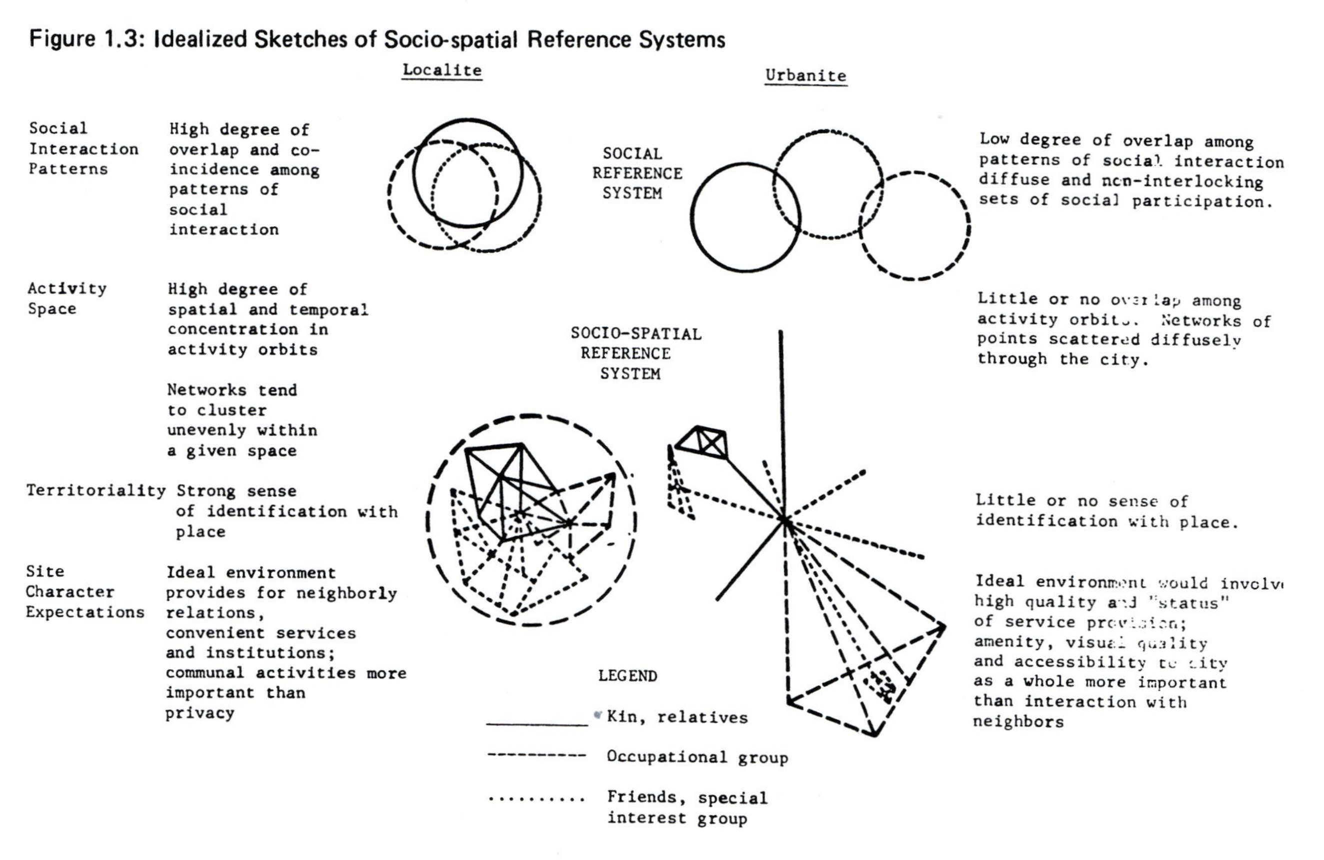

Figure 1.4: Idealized Activity Space Profiles

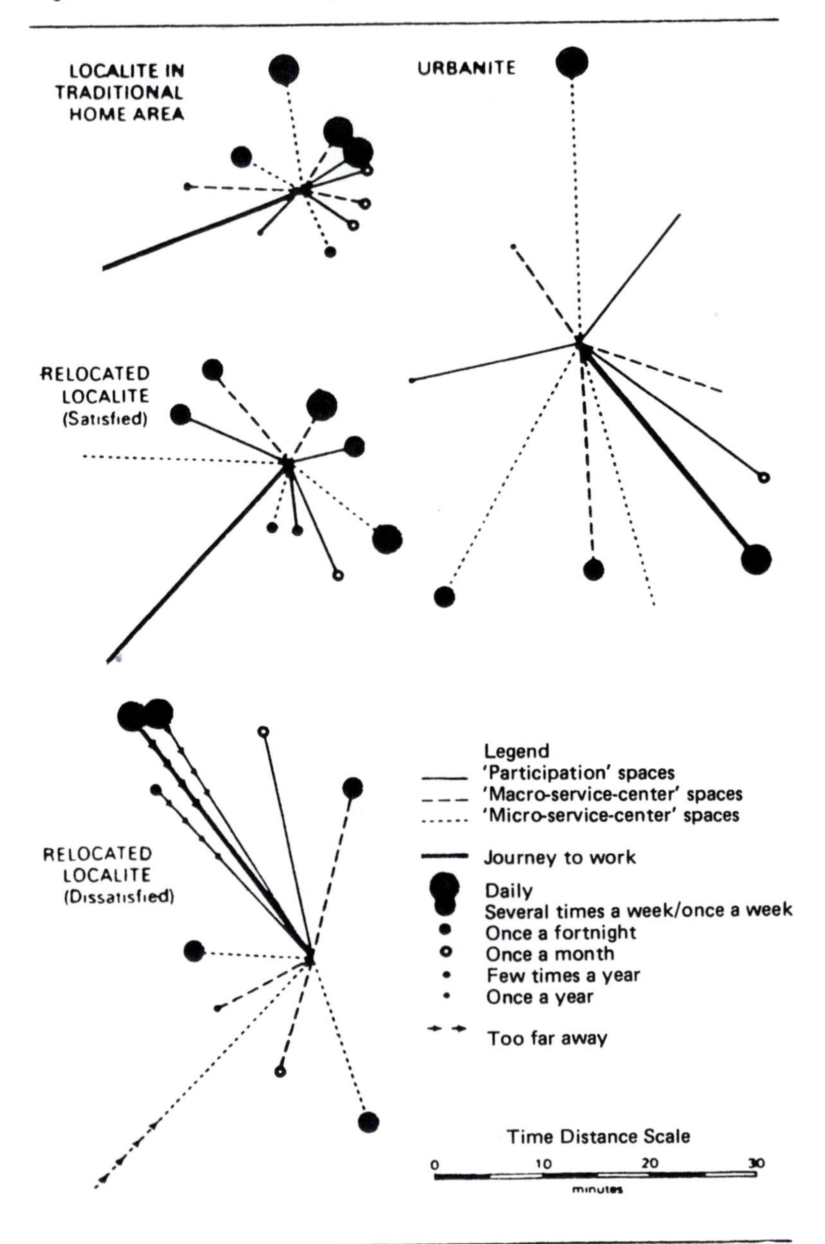

analysis of spatial and temporal concentration in movement patterns was expected to yield a horizontal zonation of socially significant spaces: (1) a local zone defined in terms of trips to shops, schools, play areas, and of casual but frequent interaction with neighbors; (2) an intermediate zone defined in terms of regular trips to occasional shops, church, doctor's office and of visits to friends, relatives, and special interest group meetings; and (3) a more diffuse zone defined in terms of interaction with close friends and relatives of primary importance even when their residential location makes visiting difficult. The ideal situation in terms of planning would be one in which the first of these zones would correspond with facilities within a five-minute orbit from home, the second would correspond to facilities located within a ten-minute orbit and the spatially more discontinuous outer zone would accommodate longer movements through space in search of a higher intensity of social meaning.

Spatial and structural overlap among these three kinds of spaces was expected to be greatest among 'localites' and least among 'urbanites' (Figure 1.4). Hence, territorial identification would be greater among the former than among the latter. It was also expected that high levels of spatial and temporal concentration in activity spaces would be associated with a propensity to value the social characteristics and micro-service features of the local environment, while diffuse patterns of activity would involve greater concern for neighborhood visual quality and 'status' and for accessibility to the city as a whole, but would place little emphasis on local neighborhood interaction. An analysis of images and associated behavior patterns could yield a typology of expectations both for local design (site) and for accessibility (situation), ranging from the polar positions of localite to urbanite (Webber, 1964). Such a typology might not parallel socioeconomic status or social class (Gerson, 1972), but there is some evidence that local network interaction contributes more to overall satisfaction with area among working-class families than among others (Yancey, 1971).

The general hypothesis outlined in Figure 1.4 suggests a restricted, roughly circular profile for the typical localite, with most daily and weekly destinations except the journey to work concentrated in the five- and ten-minute zones. The urbanite's profile, by contrast, is highly diffuse, involving little interaction with the zone closest to home. A relocated localite might be reasonably satisfied if his new activity space profile remained somewhat restricted, with frequently visited destinations still no more than ten to fifteen minutes from home. But, if the profile is greatly altered, diffuse in space and directionally biased, he will be dissatisfied.

Figure 1.5: Standard Deviational Ellipses Describing Activity Space Profiles within the Four Estates

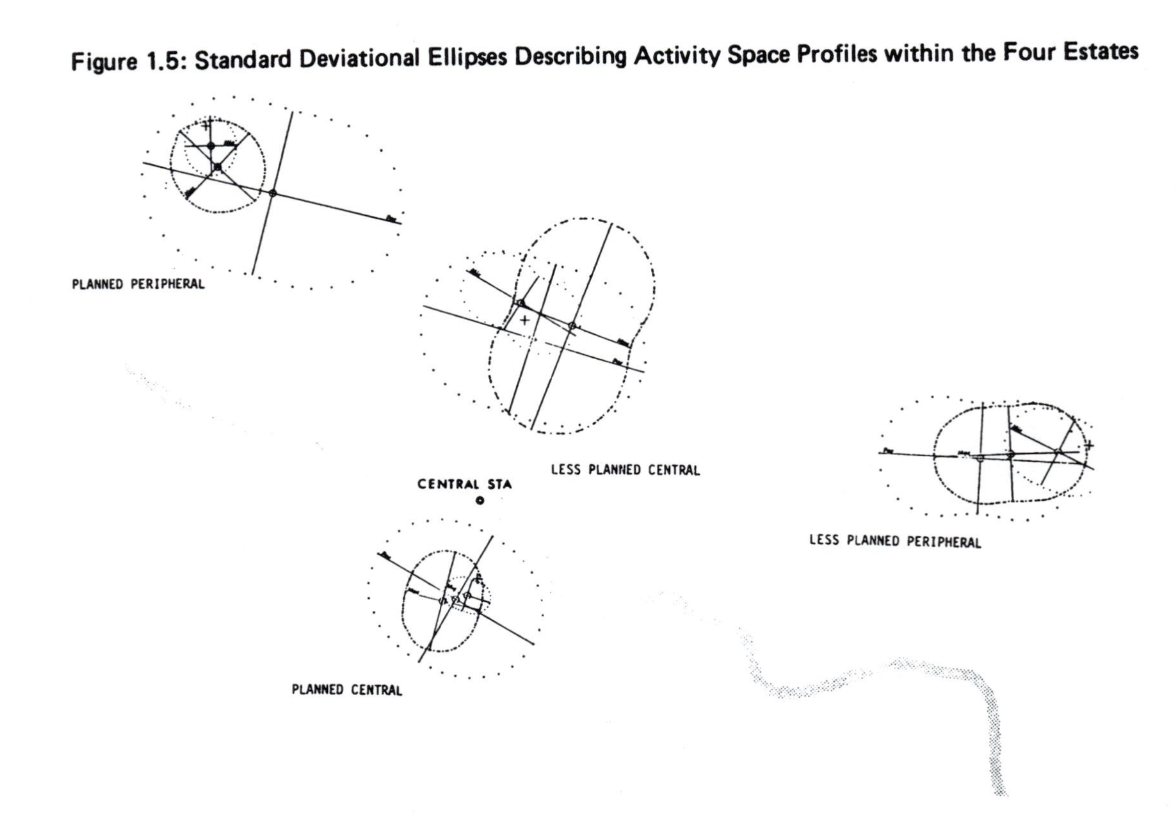

A centrographic technique known as the Standard Deviational Ellipse (Caprio, 1969; Hyland, 1970) made it possible to describe several dimensions of the aggregate activity space orbits of each resident group.[2] Ellipses (Figure 1.5) provide a graphic description of

(1) the overall volume of interaction as defined by the area of each ellipse;
(2) the degree of spatial concentration as expressed in the dimensions of minor and major axes;
(3) the general shape of the distribution expressed as a coefficient of circularity, dividing the minor axis by the major;
(4) the directional bias as indicated by the tilt of the major axis;
(5) the nature of activity space orbits, with separate ellipses for participation spaces, macro-service spaces; and micro-service spaces.

Such indices of activity space orbits form a good basis for comparing districts (that is, planned against less-planned), population sectors (households with or without children), and different territorial orientations (people who could or could not define a home area).

Once the idealized socio-spatial reference system was developed, two major avenues of research could be pursued: (1) the congruence perceived between each of the three levels of spatial experience (activity orbits, territorial identification, and image formation) and overall satisfaction with residential environment; and (2) an overall social space profile for each district based on relationships among the three levels of spatial experience.

The first line of research yielded promising clues about the appropriateness of different residential designs. For example, residents who thought of destinations as near enough were on the whole more satisfied than those who considered them too far away. Those who demonstrated a high degree of territorial identification[3] were more satisfied than those who appeared not to identify with their area. Those whose expectations about site character were realized[4] evinced greater satisfaction than those who felt that their immediate environment did not measure up to their ideal. Sectors of the population varied, to be sure, in their ranking of the importance of destinations (Figure 1.6), of neighborhood identity and of various site characteristics (Figure 1.7), and the differences have implications for planning.

The main thrust of this essay, however, is toward the second question: is there any group consistency in experience of residential environment? Can one discern any consistent pattern of association among the

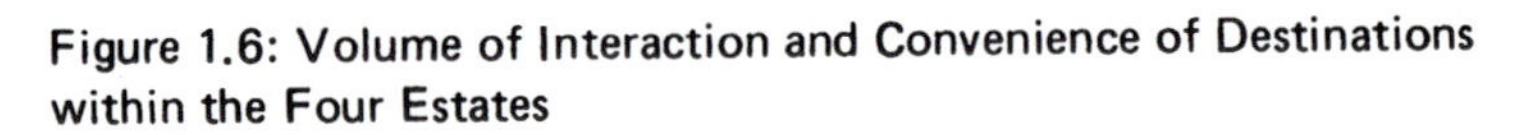

Figure 1.6: Volume of Interaction and Convenience of Destinations within the Four Estates

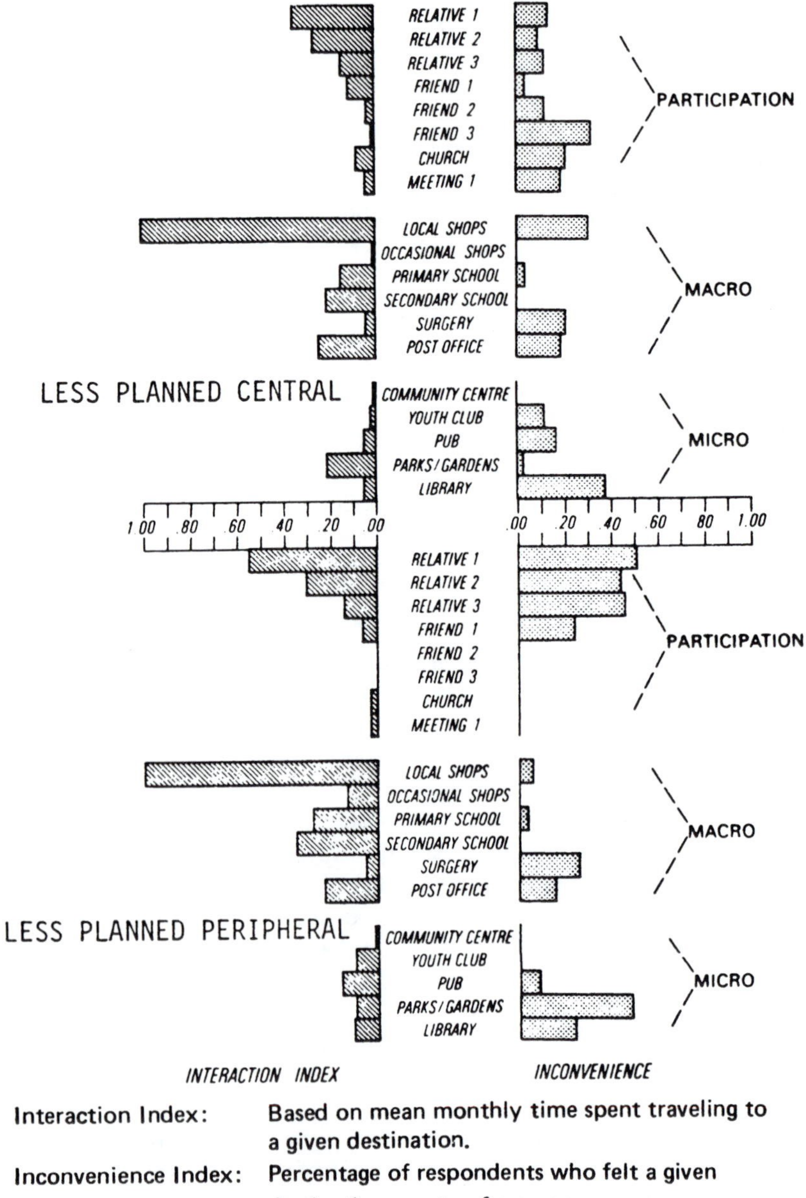

Interaction Index: Based on mean monthly time spent traveling to a given destination.

Inconvenience Index: Percentage of respondents who felt a given destination was too far away.

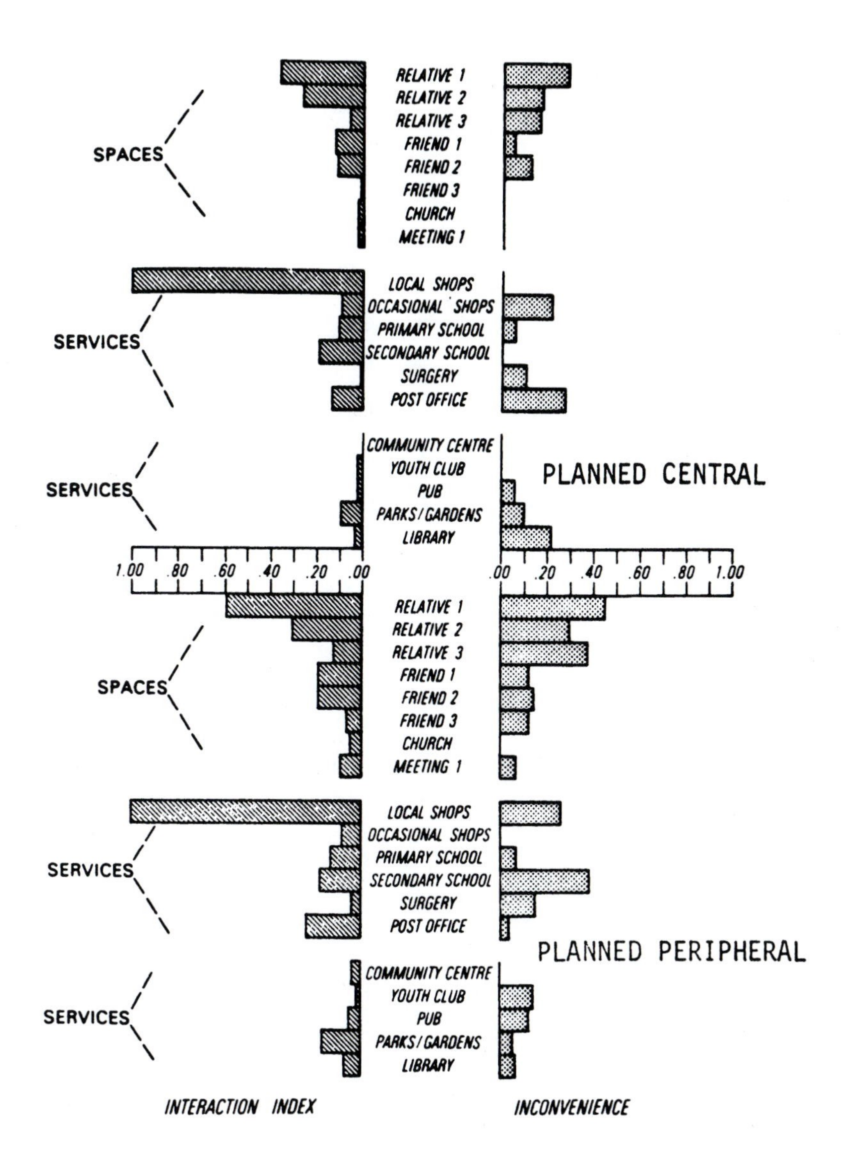
SPACES
RELATIVE 1
RELATIVE 2
RELATIVE 3
FRIEND 1
FRIEND 2
FRIEND 3
CHURCH
MEETING 1
SERVICES
LOCAL SHOPS
OCCASIONAL SHOPS
PRIMARY SCHOOL
SECONDARY SCHOOL
SURGERY
POST OFFICE
SERVICES
COMMUNITY CENTRE
YOUTH CLUB
PUB
PARKS / GARDENS
LIBRARY
PLANNED CENTRAL
1.00 .80 .60 .40 .20 .00
.00 .20 .40 .60 .80 1.00
SPACES
RELATIVE 1
RELATIVE 2
RELATIVE 3
FRIEND 1
FRIEND 2
FRIEND 3
CHURCH
MEETING 1
SERVICES
LOCAL SHOPS
OCCASIONAL SHOPS
PRIMARY SCHOOL
SECONDARY SCHOOL
SURGERY
POST OFFICE
PLANNED PERIPHERAL
SERVICES
COMMUNITY CENTRE
YOUTH CLUB
PUB
PARKS / GARDENS
LIBRARY
INTERACTION INDEX
INCONVENIENCE

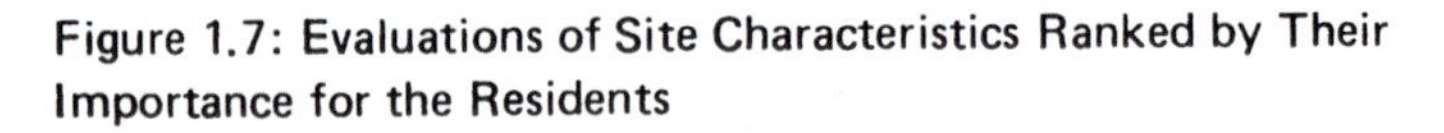

Figure 1.7: Evaluations of Site Characteristics Ranked by Their Importance for the Residents

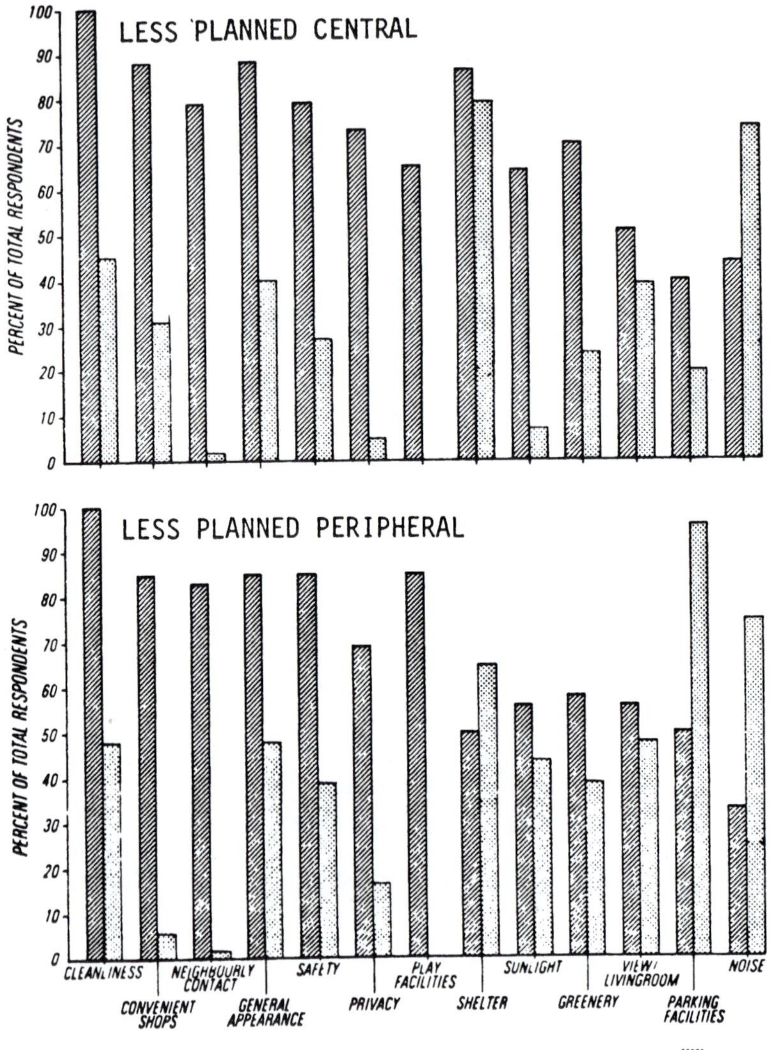

Index of Importance: Percentage of respondents who considered a particular feature 'very important.'

Index of Inconvenience: Percentage of respondents who found a particular feature less than 'very satisfactory.'

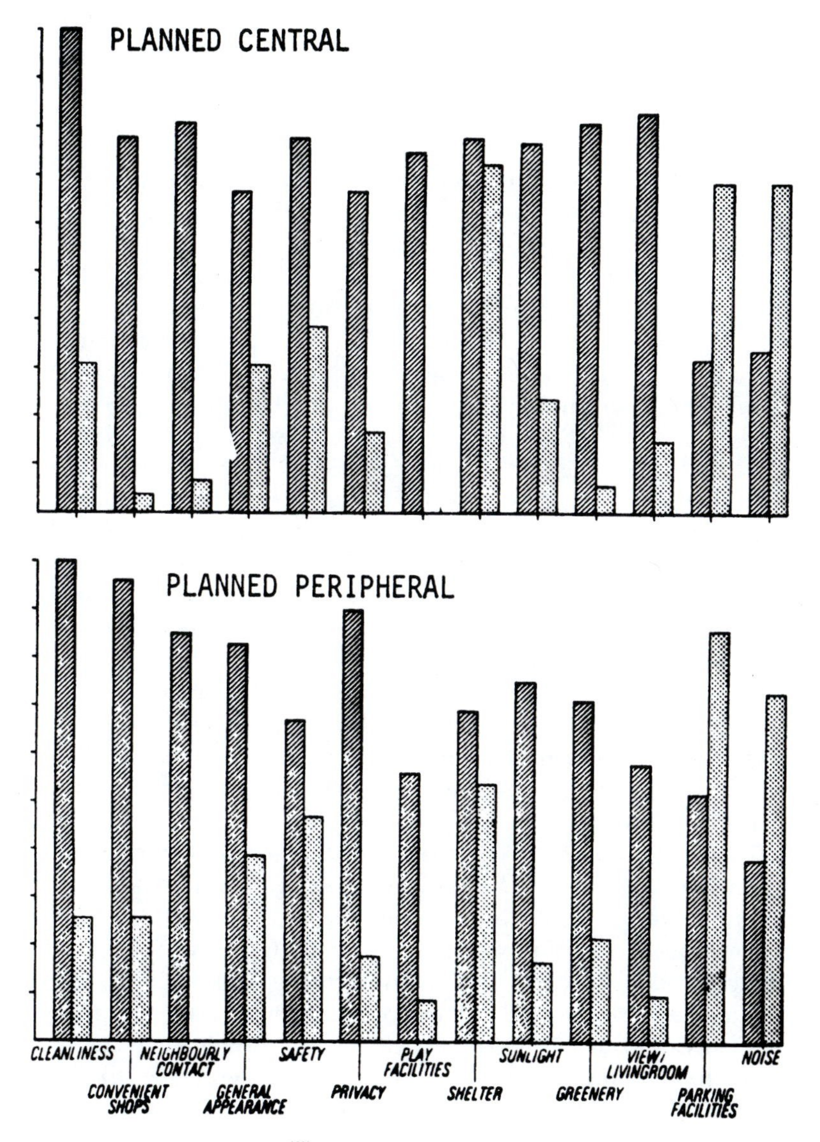
PLANNED CENTRAL
PLANNED PERIPHERAL
CLEANLINESS
CONVENIENT SHOPS
NEIGHBOURLY CONTACT
GENERAL APPEARANCE
SAFETY
PRIVACY
PLAY FACILITIES
SHELTER
SUNLIGHT
GREENERY
VIEW/ LIVINGROOM
PARKING FACILITIES
NOISE
INCONVENIENCE

territorial, interaction, and image characteristics of people within given districts? To see how resident groups have endeavored to integrate these different levels, it will be useful to describe briefly the characteristics of individual districts as a whole.

The Centrally Located Planned Estate

Of all the districts studied, this one has the most integrated pattern of interaction (Figure 1.5). The ellipse describing micro-service-center interaction is not annular but lies within 0.5 TDUs (time distance units = 10 minutes) of the zero/zero coordinate. The slight displacement of its mean center toward the southwest reflects two characteristics of the site: a physical barrier to the north (the River Clyde), and a concentration of services slightly south of the residents' homes. The same characteristics influence the shape of the ellipse describing macro-service-center interaction, but trips for occasional shopping and to secondary school may explain the north-south orientation of the macro-service ellipse. The ellipse describing participation interaction is slightly larger: friends, relatives, and kin are distributed more extensively. But the critical point is that the shift of its mean center away from the zero/zero coordinate is not great. All mean ellipse centers within this estate are located at an average distance of 0.6 TDUs from the original zero/zero coordinates. The relatively integrated nature of this estate's ellipses is reflected in the residents' high degree of territorial identification. Of the 36 respondents, 83 percent always felt at home, 75 percent could define a circumscribable home area, 56 percent had very much wanted to move into their present estate, and only 22 percent expressed any desire to leave the area.

The priorities attached to various features of the local site reveal cleanliness (1.0), view from living-room (0.83), neighborly contact (0.81), and greenery (0.81) to be the most important.[5] Fortunately, residents appeared satisfied with these features (Figure 1.7). Such features as freedom from noise (0.44), general appearance (0.67), and play facilities for children (0.42) were not considered as important. A good neighborhood was more important than a good house to 68 percent of the housewives. The fact that they found features they felt important to be satisfactory no doubt contributed to the residents' overall satisfaction with the environment. Of the inhabitants of all four estates, these people were the happiest: 64 percent here were 'very satisfied' with their area, 58 percent were 'very satisfied' with life in the area. Although no statistical evidence is yet available, it is highly likely that congruence among the three levels of environmental experience contributed to this overall sense of satisfaction with the residential environment.

The Centrally Located Less-Planned Estate

The obvious directional bias and the considerable displacements of all mean centers of interaction ellipses away from the zero/zero coordinates for this estate (Figure 1.5) provide a stark contrast with the integrated nesting of ellipses in the other centrally located estate. The areas enclosed within the macro- and micro-service ellipses for the less-planned estate are the largest in the entire sample. The macro-service ellipse is particularly striking: it is twice the size of the next largest, that of the peripheral, less-planned estate. This largely reflects trips to a denominational secondary school south of the Clyde, to churches (south of the estate), and to occasional shops (in the city center). Lacking convenience or shopping services on the estate itself, the housewives had to travel (often on foot, since bus service is not convenient) to shops on a main arterial road northeast of the estate. The mean centers of the three distributions in this estate encircle the original zero/zero coordinates. Whereas shifts of mean centers are unidirectional for the other estates, here each one shifts in a different direction. Thus, when the three types of interaction are viewed together, the resultant ellipse has an annular form.

Because this is a part of Glasgow that many of these housewives consider their own, they display some sense of territorial identification: 53 percent of the 44 sampled could define a circumscribable 'home area,' and familiar landmarks such as churches, railroad yards, even the cemetery, contributed to the feeling of belonging. Moreover, 73 percent always 'felt at home' in the district, 53 percent had wanted to move into the area, and only 26 percent wanted to leave. Even among those who were less than very satisfied with the area, only 40 percent desired to move away. Site characteristics considered 'very important' were cleanliness (1.0), general appearance (0.88), convenient shops (0.88), and shelter (0.86). That 'view from living-room' was not emphasized may reflect the fact that windows were too high up for most people (Figure 1.7).

Most of the features residents considered important, however, they did not find satisfactory. They confront a peculiarly squalid landscape, with large-scale railroad and engineering works of the late nineteenth century, a cemetery, and a generally grey-black industrial landscape on three sides. A passenger train passes beneath their windows twice every hour. Concrete, noise from traffic, and the lack of greenery or play space characterize the immediate environs. Unsatisfactory bus service, elevators nearly always out of order, rifled telephone kiosks, and a fear of 'rowdies' make life in the estate peculiarly hazardous and unsatisfactory for many residents, particularly the elderly. Why, then, the relatively high degree of satisfaction with life in the area? Whereas only 40 percent of the residents were 'very satisfied' with the area, 58 percent were 'very satisfied' with life in the area. Could it be that familiarity with the surroundings, a sense of belonging to the locality, compensates for the lack of amenities?

But the relatively high degree of satisfaction derived from this sense of belonging to the place should not be used as a justification for neglecting the specific sources of strain noted above. The provision of macro-services within a ten-minute radius of the estate would lend greater cohesion to residents' service-center orbits. Better transportation facilities might not reduce the spread of participation spaces, but they could reduce the strain for those who feel their participation destinations are too far away. This district demonstrates the need to integrate the planning of residential areas with that of the city as a whole. Many of the sources of strain in this estate emanate from city-wide functions, such as rail transport, job locations, arterial routes, and school locations. Located close to the center, the estate pays for city 'efficiency' while reaping few of its benefits.

The Planned Peripheral Estate

The provision of 'standard' services is apparent in the micro- and macro-service ellipses for this estate (Figure 1.5). The mean center of the micro-service ellipses (maximum axis 0.55; minimum axis 0.5 TDUs) is slightly displaced toward the south, reflecting the location of the new shopping center on the estate. The southerly displacement of the mean center of the macro-service ellipse is explicable in terms of occasional shopping downtown, visits to doctor, friends, relatives, and special interest groups in northwestern Glasgow, from which most of the residents came. In terms of planning standards, it is interesting to see that these service ellipses correspond with the outer limits of the ideal time-distance for the services; only the slight displacement of their mean centers detracts from the success they reflect in minimizing time traveled.

Participation shows a definite linear trend (coefficient of circularity 0.65) with the major axis inclined toward the central business district. This reflects the visits to kin and special interest groups in northwestern Glasgow, however, which are not necessarily trips to the CBD. The area of this ellipse is the largest for that class in any estate. This may reflect the location of traditional kinship ties or the willingness of upwardly mobile families to travel longer distances for special interest groups.

Of the residents of this estate, 45 percent (n=48) felt that their closest relative was 'too far away.' Given relatives' high priority ranking among all social interactions,[6] the inconvenience of reaching them presumably induced significant strain, but the length of residence in the area (87 percent had been there more than three years) and the convenience of most services seemingly had led to a sense of territorial identification: 66 percent could define a circumscribable 'home area,' 73 percent always 'felt at home' in the area. In fact, among those who were less than very satisfied with the area, only 40 percent expressed a desire to leave it. The similarity between these dispositions and those of the residents in the centrally located less-planned estate suggests some kind of trade-off between planning amenities and sense of territorial identification.

The site features most important to residents of this estate were cleanliness (1.0), convenient shops (0.96), and privacy (0.90) – characteristics generally considered important by upwardly mobile suburban populations (Gans, 1959). Such social considerations as neighborly contact also ranked high (0.85), but esthetic features were most frequently remarked on. In their free responses, residents said they like the area for its clean, healthy, and open atmosphere: 'It's good for the children, away from the noise and congestion of the city.' The high degree of satisfaction with the area (81 percent) and with life in the area (64 percent) reflects the congruence between residential aspiration and achievement. The site features considered most important were by and large satisfactory to residents. Only the difficulty of access to friends and relatives appeared to induce some strain.

The Less-Planned Peripheral Estate

The micro-service ellipse of this estate contrasts sharply with that for the planned peripheral estate (Figure 1.5). The area enclosed within this ellipse is almost three times as large as the other; the displacement of its mean center is 50 percent greater. The macro-service ellipse is almost twice the size of its planned counterpart, and the displacement of the mean center is almost twice as great. By contrast with the roughly annular tendency of service ellipses in the planned peripheral estate, the ellipses here are linear. This estate's participation ellipse is rather small; only half that of the planned peripheral estate and definitely oriented toward the CBD. Again, however, this need not reflect trips to the CBD itself. Instead, the directional bias probably reflects visits to that section of Glasgow, slightly east of the city center, from which many residents came. Especially important are visits to celebrated football grounds (Ibrox, southwest of the city center, and Parkhead, east of the city center) and to special interest groups associated with these grounds.

The physical and social isolation of this estate is reflected in residents' relatively low degree of territorial identification. Only 50 percent (n=48) said they 'always felt at home' or could define a circumscribable 'home area'; only 23 percent had very much wanted to come into the area, and 65 percent very much wanted to leave it. And of those who were less than 'very satisfied,' 80 percent very much wanted to leave.

Estate features considered important were by and large those that housewives felt were lacking in their immediate environs: cleanliness (1.0), convenient shops (0.85), general appearance (0.85), safety (0.85), and play facilities for children (0.85). In terms of most of these critical criteria as well as others – privacy (0.69), view from living-room (0.56), sunlight in all rooms (0.56), shelter (0.50), and freedom from noise (0.33) – the estate failed to meet residents' aspirations.

Toward an Integrated Perspective on Spatial Experience

At each level or mode of spatial experience, one finds a set of clues to design appropriateness. Yet there is no common metric, no coordinate system that can accommodate all the processes involved. Data for basic activity patterns – Cassirer's (1944: 42-7) 'organic level of spatial experience' – have been analyzed in a Cartesian coordinate system, but how is one to relate this to perceptual experiences of space (Michelson, 1966; Harvey, 1970)? Frequently visited places and localities of high social significance apparently stand out; points never visited, though nearer home, fade into insignificance. One appears to be dealing with a topological field, expanding and contracting according to a person's perceptual memory, normal orbits of movement, and ideas about places. And, in considering images – those abstractions about places and space that people consciously or unconsciously construct – one is dealing with a symbolic level of spatial experience that demands a still different metric and coordinate system.

If one's main purpose is to unravel the dynamics of discrete processes involved in different levels of spatial experience, then metrical coordination is a gargantuan task. But if one believes that spatial experience in daily life may be perceived as a whole, then the focus of attention

Figure 1.8: An Operational Model of Social Space

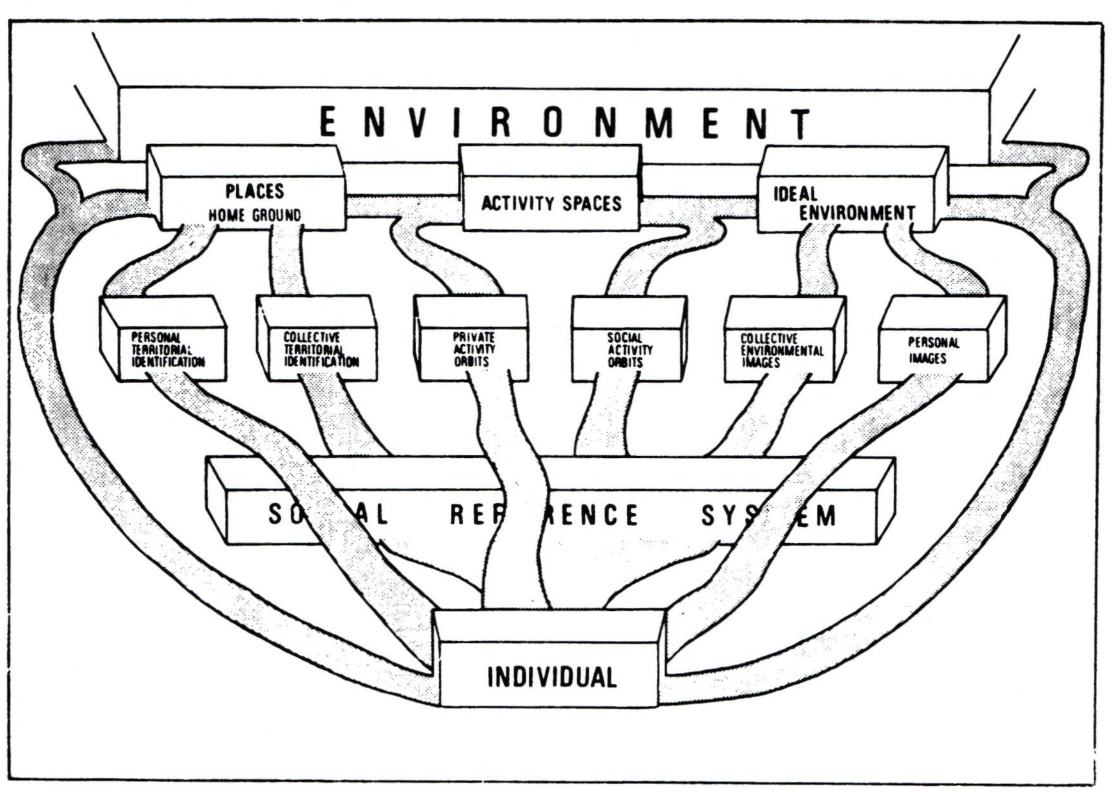

changes. One poses the question of how social reference systems, collective memories, and customary forms of interaction are expressed in the way groups assign a common meaning to space. One is concerned with the felt nature of 'experienced' spaces – for example, 'safe' and 'dangerous' places, sacred and secular spaces, focuses of social activity, highly valued zones that each group defines in its own appropriate style (Suttles, 1968; Strauss, 1961).

The main conceptual and technical points made in this chapter are summarized in Figure 1.8, a model for an integrated perspective on environmental experience. The social reference system is seen as the critical measure of significance and meaning on several dimensions of environmental experience. Whereas the Cartesian approach to measuring discrete processes seeks order and generalization, the existential view suggested here seeks meaning, specifically that ascribed to various kinds of order surrounding residential environments. Such meanings may be uniquely defined by households, but social reference systems may project collective social meanings. Such integration is evident in the consistency of the externally manifest behavior and the free-answer responses of our Glasgow residents.

Conclusion

What hypotheses about spatial behavior can be derived from this pilot study? What implications are suggested for the planning of residential areas and the provision of social services? More generally, what can the existential perspective contribute to our understanding of cities?

It may be useful first to specify the inferences that cannot justifiably be derived from the study. One is the suggestion that people in dilapidated sections of our industrial cities should be ignored, 'saved' from the redevelopment bulldozer, simply because they appear content with their traditional and cohesive life-style. Some studies, glorifying the esthetic and anthropologically exotic characteristics of slum communities, have taken redevelopment and renewal planners to task as arch-villains bent on destroying the beauty of a viable socio-spatial order (see Jacobs, 1961). While large-scale 'bulldozer' renewal programs can certainly be charged with insensitivity and have perhaps occasioned more problems than they have sought to eliminate, it is nonetheless inevitable that there be public intervention in the construction and design of residential areas. The redevelopment and renewal of slum areas, however, demands the joint response of both residents and external systems. This study attempted to isolate certain criteria and guidelines for a more productive dialogue between people and plans, between residents and planners of housing developments.

It is also a false inference that people vary so in their aspirations, tastes, and perceptions of the ideal residential environment, as to make it impossible to derive generalizations for planning. While the existential lived experience of individual families may ultimately be unique, certain consistencies in their overtly expressed attitudes and behavior patterns nonetheless contribute to their overall satisfaction with their environments. The example of 'at homeness' may illustrate this point. Among the variables associated with residential environmental satisfaction, the most significant indicated a sense of belonging to the area. The components of this phenomenon vary among districts, families, and social strata, but the phenomenon itself (an existential variable) clearly deserves further exploration. Within the small sample studied, certain measurable variables – duration of residence, location of district, stage in the family cycle, type of social interaction – appeared to be consistently related to the subjective experience of 'at homeness.' Place identification within the centrally located planned estate, for example, accompanied a high index of interaction among friends and relatives, whereas place identification in a peripheral estate accompanied a relatively higher index of interaction with service centers. This suggests that people who remain near their traditional homes continue to base their feelings of 'at homeness' on proximity to kin and friends, whereas those on the periphery require the interaction generated around services to feel 'at home.' While duration of residence in the peripheral estates is strongly related to the presence of 'home area,' this is far more evident in the planned than in the less-planned estate. This is certainly an argument for the provision of services, particularly in areas where residents have been moved far from their traditional homes. It has often been argued that residents of whole blocks or streets be moved together (Young and Willmott, 1957). Recent experiences have shown, however, that over time people are capable of building social networks in new housing estates (Young and Willmott, 1963). As suggested in a development plan for Middlesbrough (Glass, 1948), services such as schools, shops, and post offices could become the catalysts for social interaction if strategically located within a redevelopment area.

Yet it is not so much the quantity and spacing of services that raises fundamental issues as the quality, scale, and social relevance of these services. Several instances could be cited from our study in which services were said to be conveniently located, but which were the main focus of complaint in open-ended responses. Often, too, services were said to be convenient though seldom, if ever, used. In the case of doctors' offices and places of worship, for example, people sometimes preferred to travel longer distances in order to maintain contact with familiar doctors, teachers, and religious communities. Familiarity, ease of social interaction, and a variety of other factors appear to weigh quite as significantly as physical access in residents' evaluations about services.

Does this imply better transportation facilities so that people can continue to travel to familiar shops, clinics, churches, and friends and a moratorium on service provision within new residential areas? A good case could be made for transport facilities to traditional social destinations (for example, relatives and special interest groups), but a similar logic cannot in the long term be applied to services. Here the distinction between macro- and micro-services is critical. While most communities need a planned network of schools, shops, post offices, and health facilities, they vary widely in their needs and tastes regarding such micro-services as youth clubs, community centers, play areas, libraries, and nursery schools. The former should be considered mandatory from the start, the latter could be added as people grow used to one another and to their new environments.

The central implication of our study is this: the success of a residential development is contingent on the existential meaning it acquires for its residents. Who better than they can derive and infuse meaning into an environment? As far as possible, then, decisions concerning building design specifications and the range and quality of services should emerge from an active dialogue between order and meaning, the rational spatial order of technological, economic, and architectural standards and the growing self-awareness of new communities. For demand cannot be considered solely in terms of the traditional life-styles of stable communities in the slums. This would yield a set of standards which might be quite as inflexible, quite as insensitive, as standards based on the efficiencies of supply. On the demand side, one has to consider a dynamic, changing surface, evolving as people move spatially, socially, and communally. Only when a particular area design has acquired social meaning, only when its neighborhoods and physiognomy are stamped with the character of its residents and its service facilities are attuned to their needs does an ecological harmony between people and milieu emerge (Sorre, 1957; Chombart de Lauwe, 1956: 61; Gerson, 1970; Jacobs, 1961).

The problem ultimately becomes one of education both for suppliers (politicians, planners, architects) and demanders (residents). The supply structure should not be regarded as a prefabricated network of physical provisioning rationally allocated according to the constraints of technological efficiency, scale economies, and market-area potentials. It should be regarded as a potential supply system, a potential to be tapped and molded by the consciously articulated demands of resident communities.

How to enable communities to grow and develop to the point of appreciating and claiming their rights and responsibilities within the framework of the urban system as a whole is, of course, the critical problem. In a civilization imbued with the values of individualism, however, is it not likely that people feel responsibility primarily for environments they have helped create, for services they have helped organize? Is it not conceivable, then, that the collective challenge of designing and provisioning their environments could become a learning experience, generating a sense of community responsibility and contributing to identification with place? Why have we been so hesitant to experiment at different stages of the relocation process?

Why not, for example, in some areas subject to renewal, present the range of available choice a year or so beforehand to all those who are about to be moved? Ideas could be exchanged and action initiated upon consultation and collective decision. Broad decisions of building style and layout could be based on specific sociodemographic characteristics of the population, and a basic supply of macro-services built in prior to occupancy.

Through preparatory dialogue and interaction, a sense of collective community consciousness might emerge. After relocation, families would still be able to count on the support and challenge of a preexisting social order, and the business of finishing the estate design could be confronted collectively. The number, quality, and range of community centers, clubs, gardens, swimming pools, and the like could be decided as need arises and as budgetary and other constraints allow. This should prevent the appalling redundancy evident in many planned environments, where empty community centers and looted 'youth clubs' offer glaring evidence of inappropriate provision.

Such experimentation would counteract the conventional model of fitting a population into an environment prefabricated on the basis of technological, political, and economic constraints. But it would also check the utopian or Promethean model of having everyone choose his own house style and location. It calls for education toward responsible community appropriation of the rights and responsibilities for the design of environments within the context of the urban system as a whole.

The value of this model as a framework for dialogue with planners depends on refinement of analytical techniques and on comparative studies in other contexts. It provides an alternative perspective on the planning process, taking into consideration socially determined influences on demand as well as the efficiencies of supply. The value of the concept as a framework for interdisciplinary research on environmental behavior, however, poses unresolved conceptual and technical problems. The aim of this exploratory investigation in Glasgow was to open a dialogue between social scientist and planner to provide a framework in which each could contribute his unique disciplinary expertise in a climate of mutual concern over a planning dilemma. It has to some extent achieved this objective and has also pointed to ways in which research efforts on the behavioral implications of building design can be coordinated.

'The way to get at what goes on in the seemingly mysterious and perverse behavior of cities,' wrote Jane Jacobs (1961: 8), 'is to look closely, and with as little previous expectation as is possible, at the most ordinary scenes and events, and attempt to see what they mean and whether any threads of principle emerge among them.' Particularly in residential areas, the 'ordinary scenes and events' should be the primary criteria for defining design appropriateness, yet so often they have either been ignored or dismissed as extraordinary by social scientists and planners.

To discover the principles that underlie ordinary behavior in urban social worlds, and then to design a physical framework to accommodate it, are the challenge and the hope of the seventies. It is scarcely conceivable that this can be achieved without a sensitive and more unified perspective on the varying life-styles of urban communities. Planners lacking an integrated understanding of the existential character of urban life cannot be expected to design and organize its physical shell. Joint exploration of the issues involved, however, and the increased participation of citizens in the planning process itself, may enable communities, in Aristotle's phrase, to 'not merely come together in cities to live, but to stay to live the good life.'

Notes

1. This method was used for comparability with other studies of this nature (see Davis and Roizen, 1970) despite the limitations of verbally expressed 'satisfactions' as indicators of people's relationships to their environments. However, satisfaction with life in the area is a potentially better indicator of livability than conventional measures of satisfaction with dwellings and with the physical characteristics of the area.

2. The employment of geostatistical techniques in recent geographical research has demonstrated the value of the standard deviational ellipse as a measure of areal distributions (Smith *et al.*, 1968; Caprio, 1969; Hyland, 1970). The mean center of the distribution is located at the intersection of the means of the x and y axes of an arbitrarily placed Cartesian grid. Orthogonal axes are constructed through this point, and the deviations from the y axis are calculated for each location in the distribution. The standard deviation is plotted along the appropriate x axis (positive and negative). The axes are rotated through the degree 0 (usually 5° or 10°), and the procedure is repeated until the axes have been rotated through a full circle. The tracing of the standard deviations along the rotating orthogonal axes produces the standard deviational ellipse.

From this one technique, there is a yield of several comparative quantitative measures. For an areal distribution, the mean center provides a measure of the central tendency; the orientation of the major axis of the ellipse shows the locational trend; the ratio of the minor to the major axis of the ellipse gives an index of circularity, while the shape and area of the ellipse are further indicators of the dispersion. This technique has generally been applied to the distribution of data in geodesic space, but our data on individuals' perceptions of time-distance were non-geodesic. Variations in perceived distance were assumed to be either minimal or normally distributed. Ellipses were used primarily for descriptive rather than analytical purposes. (The writer is indebted to Gerard Hyland for this analysis of activity spaces.)

3. Territorial identification was measured in terms of (a) feeling at home, (b) ability to define a home area, and (c) desire to move into the area or no desire to move away from the area, and (d) duration of residence in the area.

4. Various design characteristics were ranked in terms of their perceived importance and then evaluated by the residents. Figure 1.7 graphically places the evaluation of site characteristics in the context of respondents' images of ideal residential environments.

5. The 'Index of Importance' for each feature was based on the percentage of the population who considered it to be 'very important'; for example, 0.81 for 'greenery' means that 81 percent of the respondents considered that feature very important.

6. Destinations were ranked within each estate in terms of an interaction index based on mean monthly time spent traveling to that destination.

References

Adams, J.S. 1969. 'Directional Bias in Intra-urban Migration,' *Economic Geography, 45,* 302-23.

Alonso, W. 1964. *Location and Land Use.* Cambridge, Mass.: Harvard University Press.

Alpaugh, D. (ed.), 1970. *Design and Community.* Raleigh: North Carolina State University School of Design.

Altman, I. 1970. 'Territorial Behavior in Humans: an Analysis of the Concept.' In L.A. Pastalan and D.H. Carson (eds.), *Spatial Behavior of Older People*, pp. 1-24. Ann Arbor, Mich.: Wayne State University Institute of Gerontology.

— and Haythorn, W.E. 1967. 'The Ecology of Isolated Groups,' *Behavioral Sci., 12,* 169-82.

Archea, J. and Eastman, C. (eds.), 1970. *EDRA 2: Proceedings of the Second Annual Environmental Design Research Association Conference.* Pittsburgh: Carnegie-Mellon University.

Ardrey, R. 1966. *The Territorial Imperative: A Personal Inquiry into the Animal Origins of Property and Nations.* New York: Atheneum.
Back, K.W. 1962. *Slums, Projects and People.* Durham, NC: Duke University Press.
Barker, R.G. 1968. *Ecological Psychology.* Stanford, Calif.: Stanford University Press.
Barth, F. 1956. 'Ecological Relationships of Ethnic Groups in Swat, North Pakistan,' *Amer. Anthropologist, 58,* 1079-89.
Bell, W. 1959. 'Social Areas: Typology of Urban Neighborhoods.' In M. B. Sussman (ed.), *Community Structure and Analysis*, pp. 61-92. New York: Thomas Y. Crowell.
Berry, B.J.L. (ed.), 1971. 'Comparative Factoral Ecology,' *Economic Geography, 47* (June), 209-33.
— and Horton, R.E. 1970. *Geographic Perspectives on Urban Systems.* Englewood Cliffs, NJ: Prentice-Hall.
Blair, T.L. 1969. 'Poverty of Urban Planning,' *Official Architecture and Planning, 32* (February), 181-6.
Blaut, J.M. and Stea, D. 1971. 'Studies of Geographic Learning,' *Annals of Association of American Geographers, 61*, 387-93.
Boal, F.W. 1969. 'Territoriality on the Shankill-Falls Divide, Belfast,' *Irish Geography*, *6*, 30-50.
Bogardus, E.S. 1925. 'Measuring Social Distance,' *J. of Applied Sociology*, *9* (January-February), 299-308.
Broady, M. 1968. *Planning for People.* London: National Council of Social Service.
Brown, L.A. and Moore, E.G. (eds.), 1971. 'Perspectives on Urban Spatial Systems,' *Economic Geography, 47* (1) (January).
Buttimer, Anne, 1969. 'Social Space in Interdisciplinary Perspective,' *Geographical Review, 59* (July), 417-26.
—, 1971. 'Sociology and Planning,' *Town Planning Review, 42* (April), 145-80.
Caprio, R.J. 1969. 'A Geostatistical Analysis of Populations and Housing Distributions: Newark, N.J., 1940-60,' MA thesis, University of Cincinnati.
Cassirer, E. 1944. *An Essay on Man.* New Haven, Conn.: Yale University Press.
Chapin, H.S. and Hightower, H.C. 1966. *Household Activity Systems: A Pilot Investigation.* Chapel Hill, NC: Institute for Research in Social Science.
Chombart de Lauwe, P.H. *et al.* 1952. *Paris et l'agglomération parisienne.* Paris: Presses Universitaires de France.
—, 1956. *La vie quotidienne des familles ouvrières.* Paris: Presses Universitaires de France.
—, 1965. *Paris: Essais de sociologie, 1952-1964.* Paris: Presses Universitaires de France.
Cox, K.R. and Golledge, R.G. (eds.), 1969. *Behavioral Problems in Geography: A Symposium.* Evanston, Ill.: Northwestern University Studies in Geography.
Cox, K.R. and Zannaras, G. 1970. 'Designative Perceptions of Macro-spaces: Concepts, a Methodology and Applications,' *Environmental Design Research Assn. Proceedings, 2,* 118-34.
Craik, K.H. 1970. 'Environmental Psychology.' In K.H. Craik *et al.*, *New Directions in Psychology*, pp. 1-121. New York: Holt, Rinehart & Winston.
Davis, G. and Roizen, R. 1970. 'Designative Perceptions of Macro-spaces: Concepts, a Methodology and Applications,' *Environmental Design Research Assn. Proceedings*, *2*, 28-44.
Downs, R.M. 1971. 'Geographic Space Perception: Past Approaches and Future Prospects.' In M. Beard *et al.* (eds.), *Progress in Geography*, pp. 65-108. London: Edward Arnold.
— and Horsfall, R. 1971. 'Methodological Approaches to Urban Cognition.' Presented at the Sixty-Seventh Meeting of the Association of American Geographers, Boston.

Flachsbart, P.G. 1969. 'Urban Territorial Behavior,' *J. of Amer. Institute of Planners, 27* (November), 412-16.
Fried, M. and Gleicher, P. 1961. 'Some Sources of Residential Satisfaction in an Urban Slum,' *J. of Amer. Institute of Planners, 27* (November), 305-15.
Gans, H.J. 1959. 'The Human Implications of Slum Clearance and Relocation,' *Journal of American Institute of Planners, 25* (February), 15-25.
—, 1961. 'Planning and Social Life: Friendship and Neighbor Relations in Suburban Communities,' *Journal of American Institute of Planners*, *27* (2), 134-40.
—, 1968. *People and Plans: Essays on Urban Problems and Solutions.* New York: Basic Books.
Gerson, E. 1972. 'Social Organization of Urban Neighborhoods,' PhD dissertation, University of Chicago.
Gerson, W. 1970. *Patterns of Urban Living*. Toronto: University of Toronto Press.
Glass, R. (ed.), 1948. *The Social Background of a Plan*. London: Routledge & Kegan Paul.
Goffman, E. 1961. *Encounters: Two Studies in the Sociology of Interaction*. New York: Bobbs-Merrill.
Greer, S. 1956. 'Urbanism Reconsidered: a Comparative Study of Local Areas in a Metropolis,' *Amer. Soc. Rev., 21,* 19-25.
—, 1969. *The Logic of Social Inquiry.* Chicago: Aldine.
Gutman, R. 1966. 'Site Planning and Social Behavior,' *J. of Social Issues, 22,* 103-15.
Hall, E.T. 1966. *The Hidden Dimension.* New York: Doubleday.
Harvey, D.N. 1970. 'Social Processes and Spatial Form: an Analysis of Conceptual Problems in Urban Planning,' *Papers of the Regional Science Assn., 25,* 47-69.
—, 1972. 'Revolutionary and Counter Revolutionary Theory in Geography and the Problem of Ghetto Formation,' *Antipode, 3* (July).
Hemmens, G.C. 1966. *The Structure of Urban Activity Linkages.* Chapel Hill, NC: Institute for Research in Social Science.
Hole, V. 1959. 'Social Effects of Planned Rehousing,' *Town Planning Rev., 30* (July), 161-73.
Hyland, G.A. 1970. 'Social Interaction and Urban Opportunity: the Appalachian In-migrant in the Cincinnati Central City,' *Antipode*, *2* (December), 68-83.
Hyman, H. and Singer, E. (eds.), 1968. *Readings in Reference Group Theory and Research.* New York: Free Press.
Jacobs, J. 1961. *The Death and Life of Great American Cities.* New York: Vintage.
Keller, S. 1969. *The Urban Neighborhood: A Sociological Perspective*, New York: Random House,
Kelly, G.A. 1958. 'Man's Constructions of his Alternatives.' In G. Lindsey (ed.), *Assessment of Human Motives,* pp. 32-64. New York: Holt & Rinehart.
Lee, T. 1968. 'Urban Neighborhood as a Socio-spatial Schema,' *Human Relations, 21,* 241-68.
Lewin, K. 1951. *Topological Field Theory in Social Science.* New York: Harper.
Lorenz, K. 1966. *Aggression.* New York: Harcourt, Brace & World.
Lynch, K. 1960. *The Image of the City.* Cambridge, Mass.: Harvard University Press.
Metton, A. 1969. 'Le quartier: étude géographique et psychosociologique,' *Canadian Geography, 13,* 299-316.
Michelson, W. 1966. 'An Empirical Analysis of Urban Environmental Preferences,' *J. of Amer. Institute of Planners, 24* (November), 355-60.
Muth, R.F. 1969. *Cities and Housing.* Chicago: University of Chicago Press.
Park, R.E. 1924. 'The Concept of Social Distance,' *J. of Applied Sociology, 8* (July/August), 339-44.

Peterson, G.L. 1967. 'A Model of Preference: Quantitative Analysis of the Perception of the Visual Appearance of Residential Neighborhoods,' *J. of Regional Sci., 7,* (1), 19-30.
Proshansky, H.M. *et al.* (eds.), 1970. *Environmental Psychology: Man and His Physical Setting.* New York: Holt, Rinehart & Winston.
Rainwater, L. 1966. 'Fear and the House-as-haven in the Lower Class,' *J. of Amer. Institute of Planners, 32* (January), 23-31.
Reade, E. 1969. 'Contradictions in Planning,' *Official Architecture and Planning, 32,* 1179-83.
Rothblatt, D.N. 1961. 'Housing and Human Needs,' *Town Planning Rev., 42,* (2), 130-44.
—, 1964. *Human Needs and Public Housing.* New York: New York City Planning and Housing Library.
Runciman, W.G. 1966. *Relative Deprivation and Social Justice: A Study of Attitudes to Social Inequality in Twentieth Century England.* Berkeley: University of California Press.
Schorr, A.L. 1963. *Slums and Social Insecurity.* Washington, D.C.: US Department of Health, Education, and Welfare.
Shevky, E. and Bell, W. 1955. 'Social Area Analysis,' *Stanford Sociological Series 1*, Palo Alto.
Shevky, E. and Williams, M. 1949. *The Social Areas of Los Angeles*, Berkeley: University of California Press.
Shibutani, T. 1955. 'Reference Groups as Perspectives,' *Amer. J. of Sociology, 60,* 562-9.
Simmie, J.M. 1968. 'Social Survey Method in Town Planning,' *J. of Town Planning Institute, 54* (May), 222-9.
Smith, R.V., Flory, S.E., Bashshur, R.L., and Shannon, G.W. 1968. 'Community Support for the Public Schools in a Large Metropolitan Area,' US Office of Education Project 2557. Ypsilanti: Eastern Michigan University.
Sommer, R. 1969. *Personal Space: The Behavioral Basis of Design.* Englewood Cliffs, NJ: Prentice-Hall.
Sorokin, P.D. 1928. *Social Mobility.* New York: Free Press.
Sorre, M. 1957. *Rencontres de la géographie et de la sociologie*. Paris: Marcel Rivière et Cie.
Stea, D., and Downs, R. (eds.), 1970. 'Cognitive Representations of Man's Spatial Environment,' *Environment and Behavior* (June).
Strauss, A. 1961. *Images of the American City.* New York: Free Press.
Suttles, G.D. 1968. *The Social Order of the Slum.* Chicago: University of Chicago Press.
Theodorson, G.A. and Theodorson, A.G. 1969. *A Modern Dictionary of Sociology.* New York: Thomas Y. Crowell.
Webber, M.W. 1964. 'Culture, Territoriality and the Elastic Mile,' *Papers of the Regional Sci. Assn., 13,* 59-69.
Wilner, D. *et al.* (eds.), 1962. *The Housing Environment and Family Life.* Baltimore, Md.: Johns Hopkins Press.
Yancey, W.L. 1971. 'Architecture, Interaction and Social Control: the Case of a Large-scale Housing Project,' *Environment and Behavior, 3,* 3-22.
Young, M., and Willmott, P. 1957. *Family and Kinship in East London.* London: Routledge & Kegan Paul.
—, 1963. *The Evolution of a Community*, London: Routledge & Kegan Paul.

2 TOWARD A GEOGRAPHY OF GROWING OLD*

Graham D. Rowles

Marie was an 83-year-old widow. She lived alone in a small brick house in a working-class inner-city neighborhood in the northeastern United States. When her husband purchased the home, 57 years before I met her, the neighborhood was a vibrant French-Canadian community. In recent years, zone-in-transition land uses had encroached, stores and social institutions had closed, and a low-income renter population had filtered into the dilapidated housing stock. Younger, more affluent households, including those of her children, had long since opted for suburban living, as the locality had become increasingly inhospitable with frequent incidents of violent crime, arson, and vandalism. Marie and most of her elderly peers, however, still dwelled in the neighborhood they knew as home.

How did Marie cope with the experience of growing old in this seemingly inhospitable setting? Had the inexorable erosion of her physical and sensory capabilities and growing alienation from the contemporary neighborhood's social milieu resulted in spatial withdrawal? How was she affected by the prospect of the neighborhood's eventual demise – the passing of this place she had known?

To my initial bewilderment, Marie seemed uninterested in these questions I had culled from extensive review and reflection on the literature on aging. She was reluctant to talk about physical restriction, reduced access to services, spending more time at home, problems of social abandonment, or fears of the future. Instead, as we sat in her parlor poring over treasured scrapbooks in which she kept a record of her life, she would animatedly describe trips she had taken to Florida many years previously. She would muse on the current activities of her granddaughter in Detroit, a thousand miles distant. She would describe incidents in the neighborhood during the early years of her residence. Blinded by preconceptions, I could not comprehend at first the richness of the taken-for-granted lifeworld she was unveiling.

The recent emergence of humanistic geography is premised on reveal-

*Portions of this essay are abstracted from a longer study, Graham D. Rowles, *Prisoners of Space? Exploring the Geographical Experience of Older People* Westview Press (Boulder, Colorado, 1978). The author wishes to thank the publisher for allowing him to abstract these portions here.

ing such lifeworlds in terms of meanings, values, and intentionalities which permeate them (Tuan, 1976; Entrikin, 1976; Ley and Samuels, 1978). An outpouring of good intention has resulted in intriguing and perceptive commentary on the manner in which individuals and groups experience the spaces and places of their lives (Relph, 1976; Tuan, 1977; Seamon, 1979). In terms of acceptance, both within and outside the discipline, much of this work is pervaded by an image of esotericism. But humanistic geography is far from esoteric. Recently, researchers have begun to harness more directly the potential of the perspective for making pragmatic contributions to theory construction, clinical practice, and the formation of public policy more sensitive to life experience (Godkin, 1977; Rowles, 1979). This essay seeks to complement the trend. I offer an interpretation of Marie's *geographical experience* – her involvement within the spaces and places of her life – and that of several of her age peers which was gradually revealed as I became more deeply immersed in the subjectivity of their lives. This interpretation is presented and assessed as a contribution to the development of a humanistic geography of growing old.[1]

An Exploration

Societal lore reinforces an empiricist image of the older person's geographical experience pervaded by a motif of closure – of progressive imprisonment within a more limited space (Pastalan, 1971, 2; Montgomery, 1977, 253). Reduced biological, psychological and social capabilities are widely considered to be accentuated by increasingly insistent environmental constraints, including barriers imposed by design of the physical setting, economic deprivations, transportation restrictions, social spatial isolation, and confining societal attitudes. The outcome, it is argued, is a progressively more limited activity orbit and, as environmental vulnerability increases, a growing concern for remaining in secure and familiar settings.

The image is clearly naive, indeed almost a parody, but is deeply ingrained in the gerontology literature and often internalized by the elderly. One reason for its durability is a grounding in studies of overt behavior patterns – the most easily monitored manifestations of geographical experience. In contrast, there is a paucity of inquiry into older people's perceptual orientation within large-scale environments, their emotional and generally prereflective identification with the places of their lives, and the way in which life experience is incorporated within the constitution of a lifeworld. A sensitivity to these levels of experience, I discovered, was essential for understanding Marie. Moreover, in a developed world where the elderly are becoming an increasingly significant vulnerable population, it is becoming ever more apparent that there is a need to develop a more sophisticated perspective on the changing relationship between the older person and his or her environmental context – one comparable to our growing awareness of the changing spatial world of the child.

To evolve preliminary insight, I undertook a three-year in-depth study of five elderly individuals. In addition to Marie, I came to know Evelyn (76 years old), Edward (80), Raymond (69), and Stan (68). All had resided in the neighborhood for over forty years. A close interpersonal relationship developed with each participant as, during frequent meetings, we mutually explored a hitherto taken-for-granted aspect of their lives. Data-gathering in a formal sense was limited. Instead, through intersubjective encounter, I sought immersion within the participants' lifeworlds and to learn through 'creative dialogue' in a process of mutual discovery (Von Eckartsberg, 1971; Rowles, 1978b). A collection of materials was assembled on each person, comprising reams of notes, photographs, and sketch maps; and many hours of taped conversations.

Making sense of the experience was an inductive process in which I made a conscious effort to transcend preconceptions and to derive insight solely from the 'text' of the encounters. I drafted a detailed descriptive vignette on each person. Using their own words, I sought to portray the subtle complexity of each participant's life-style and involvement within the spaces and places of his or her life.[2] As I worked on the vignettes, I searched for consistent shared themes in geographical experience. This involved the honing of insights emerging during our exchanges, together with a classifying and reclassifying of material from the tape recordings, over several months of post-field-work reflection. Finally, I held a series of meetings with participants to present a slowly crystallizing perspective and to solicit feedback.

Modalities of Geographical Experience

The participants' geographical experience expressed a subtle meshing of space and time, embracing not only physical and cognitive involvement within their contemporary neighborhood, but also vicarious participation in an array of displaced environments. I eventually isolated four distinct but overlapping experiential modalities – *action, orientation, feeling,* and *fantasy.*

Action

Action involved physical locomotion on three conceptually distinguishable levels. Movement within the proximate physical setting, such as reaching for a cupboard or traversing a room, was designated as immediate action. All the participants admitted to declining bodily agility. Their actions on this level exhibited increasingly efficient use of space within their homes and conservation of personal energy – manifest in the closing off of unneeded, less accessible rooms and the judicious rearrangement of furniture and appliances.

On a larger scale, everyday activity consisted of routine service, social, and recreational trips. These were generally confined to the city and seemed by their regularity to provide a time-space rhythm within the participants' lives. Actions on this level had also become more limited and selective as the participants accommodated to growing old. Trips were confined to daylight hours. Shopping was often undertaken by family or friends; the radio sermon substituted for the journey to church; visits to friends became more localized or replaced by lengthy telephone conversations.

A third class of actions, occasional trips, confounded the stereotype. In spite of the economic constraints under which they lived, the participants made occasional trips to visit their families or vacations to locations far from their homes. Evelyn, the summer before I met her, was subsidized by her family to visit a son in Arizona. 'My first time I ever flown, too!' she exclaimed. Raymond spent three months with a daughter in Arkansas and four with a son in North Dakota. Marie had taken a lengthy trip, traveling first to Florida to visit her daughter, then to Detroit to stay with her son, and finally to Quebec to revisit her birthplace. The participants had made such trips more frequently as they had grown older.[3]

Orientation

Actions were framed in terms of a cognitive differentiation of space involving *schemata* – mental representations of physical-social space providing *orientation* within a 'known' world (Tuan, 1975a; Downs and Stea, 1973, 1977; Moore and Golledge, 1976). A personal schema furnished implicit psychobiological orientation – a preconscious sense of left and right, horizontal and vertical, back and front (Bollnow, 1967, 179; Howard and Templeton, 1966). Physiological decrements associated with aging often make it difficult to maintain balance (Shephard, 1978, 134). As they had grown older, Stan and Edward had compensated by using a cane; and all five participants tended to avoid open spaces and settings lacking environmental supports.

Similar adaptation was apparent in an intimate awareness of familiar routes. Stan, a regular patron of local bars, would shuffle from one establishment to another tracing consistent paths. Sensitive to cracks in the sidewalk, he knew the paths affording shade on a hot afternoon and safety on the icy days he dreaded. He was aware of the street crossings most hazardous during lunchtime traffic. He had internalized a series of detailed 'specific' linear schemata which facilitated his routinized movement under diverse environmental conditions.

Such specific schemata were embedded within a superordinate general schema which represented a more basic differentiation of milieu into a series of experientially distinctive annular domains. Home, fulcrum of the lifeworld, was possessed inviolable space. Immediately outside, a narrow surveillance zone, encompassing the field of vision from the home, was distinguished by a sense of partial control and watchful reciprocity among neighbors. As Raymond remarked:

> I check on her when she puts her shades down. Almost every day I look out. If there's a light there and I see it, I'll wave. She knows that I'm home. And then if anything should happen, she can grab the phone and call me.

Another neighbor monitored Raymond: 'If he don't see me at night sitting down on the couch watching television, he'll come bounding over here, to see if anything's wrong.' The neighborhood was less clearly sensed as a physical space but clearly distinguished as social space, the territorial preserve of a French-Canadian community. Incorporation of the city and spaces beyond tended to be more fragmentary and amorphous. However, significant 'privileged places' such as the homes of children, former residences, and vacation resorts were known in intimate detail. They stood out as anchoring points within the landscape of the general schema.

Feeling

This experiential landscape was more than a static pictorial representation: it was a shifting collage, tinted subtly by a reservoir of *feeling* infused within individual locations and in turn evoked by them. Over the years, layer upon layer of experience became incorporated within the cognized meaning of particular settings; they emerged as affective symbols providing extensions of individual identity (Ley, 1977, 508). For Marie, the ramshackle hall where many years ago she danced in a red velvet dress on her silver wedding anniversary would always be a special place. It was a focus of personal feeling.

Rarely are meanings fully private. Places are also imbued with shared feelings. They become 'fields of care' (Tuan, 1975b, 236). Sentiments for neighborhood space, held in common and reinforced by interaction among a social network of elderly age peers, expressed a mutual sense of belonging and continuity with the community of a more auspicious past. Such feelings provided reassurance in accommodating to distressing contemporary environmental change. The supportive potential of shared affinity for place was often subtly revealed. When questioned directly, the participants could list few close friends. However, when confronted with a comprehensive listing of 455 elderly neighborhood residents, invariably at least 200 could be identified and anecdotes recounted pertaining to each one. Relationships had often developed many years previously. In some cases, unbeknown to the participants, identified individuals were no longer alive. This was immaterial, for a symbolic presence was enough to sustain a sense of belonging.

Fantasy

Perhaps the most interesting findings pertain to a fourth modality of geographical experience – *fantasy.* This term is not used in any negative, demeaning, or perjurious sense, but as a general designation for a modality of experience having particular significance within the totality of the participants' geographical experience. Recall my opening observations on Marie's reluctance to focus on what I had preconceived as her environmental experience. She was not alone. Listening to the tapes it became apparent that for long stretches of conversation, the participants' thoughts were far removed from the rooms in which we sat. Stan mused on the Poland of his boyhood; Evelyn during reflective monologues participated in the affairs of a son in Arizona; all the participants frequently reminisced on events in the neighborhood of the past. Gradually I came to appreciate the significance of such activity. Vicarious immersion in event implies vicarious immersion in place: it facilitates a transcendence of location.

Two forms of fantasy could be distinguished. *Reflective geographical fantasy* involved reminiscence, a selective participation in environments of the past. Marie could re-enter the neighborhood of 1930:

> On Imperial Street there used to be a lot of stores. There was a grocery store there, and jewelry shop, and barber's shop . . . and on the other side there was a fish market. Two grocery stores on the other side of the street. Mercier used to have a meat market at the back and grocery store at the front. They used to have a church there right on the corner of Easthill and Imperial. Father Deigneaux used to say Mass down there.

The place in which she currently dwelled was far more than the physically deteriorating contemporary setting I could view. It was a series of places through time. Each could be evoked as incidents were recalled. Marie could also reinhabit selectively reconstituted places of her childhood. Indeed, in reverie, the participants immersed themselves in settings expressing the entire time-space continuum of their personal histories.

Projective geographical fantasy transported the participants to spatially removed contemporary milieu – notably environments where their children resided. Raymond had a garden at his daughter's home in Arkansas which he 'tended' with care from a distance of many hundreds of miles. Evelyn often 'participated' in an Arizona family: savoring a graduation, sharing a concern over the inclement weather she monitored through television forecasts, musing over the everyday affairs of a distant household. This tendency for projection into the worlds of children has been noted by Hochschild, who describes a propensity for 'altruistic surrender' as the older person lives through the experience of a child and gains a sense of intimacy and identification compensating for spatial separation (Hochschild, 1973, 96-111).

Clearly, the four modalities of geographical experience are not mutually exclusive. They comprise dimensions of a total complex of 'being' within a lifeworld (summarized in Figure 2.1). Geographical experience, at least for Stan, Marie, Raymond, Evelyn, and Edward, involved more than mere behavioristic locomotion through timeless Cartesian space. Rather, it was a fusion of implicit awareness, thought, and action, entailing holistic involvement within a 'lived space' – a lifeworld with temporal depth and meaning as well as spatial extent. Each participant's geographical experience revealed internal consistency, a consonance among the modalities. In concert, these modalities expressed a current experiential state of adjustment within a dynamic person-environment transactional system.

Figure 2.1: The Elderly Person's Lifeworld

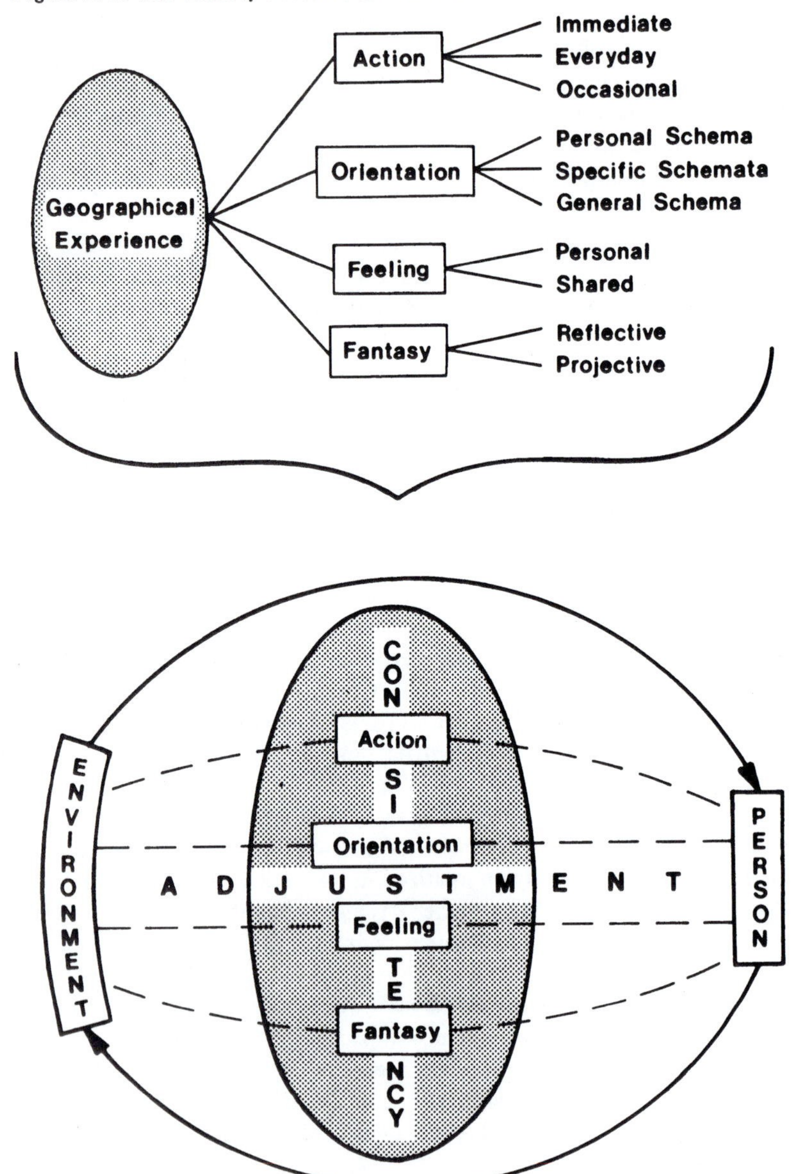

The Geographical Experience of Growing Old

The perspective developed through collaboration with Marie and her peers could, of course, mirror the experience of individuals of any age. The important question is whether it is possible to develop one stage further and to identify distinctive transitions in geographical experience associated with growing old.

Observation and lengthy discussions with the participants suggested that it is not so much the structure of geographical experience that changes as emphasis among the modalities which changes with advancing age. There is a progressive *limitation in the realm of action* accompanied by *expansion in the role of geographical fantasy*. Withdrawal from full physical environmental participation was compensated by greater emphasis upon contemplative vicarious modes of environmental participation.[5] This hypothesis is consistent with research indicating a propensity for increasing 'interiority,' reminiscence, and a process of life review among the elderly (Neugarten *et al.*, 1964; McMahon and Rhudick, 1967; Lewis, 1971; Butler, 1963; Coleman, 1974). The process expressed an accommodation to growing old which, for the participants, provided liberation from the time-space constrictions of personal decline and a deteriorating contemporary neighborhood.

This major transition was accompanied by a series of associated changes in both orientation within space and feeling about place. These reflected a selective intensification of cognitive involvement. Specific schemata for remaining paths of action had become more refined. Familiarity with space facilitated progressive reorientation as the older person adjusted to declining physical and psychological capability. Guiding landmarks and potential hazards were highlighted through 'selective attention' to salient environmental features – cracks in the sidewalk, dangerous street crossings, potential resting places, and so on. As more time was spent at home, there was an intensification in the importance of proximate zones (especially the home and surveillance zone) and – because of increasing vicarious involvement within them – in significant displaced environments within the general schema. Transitions in the participants' feelings for place included intensified affective bonds with proximate space, environments of their children, and other locations of meaning within their life histories.

In sum, as the participants had grown older, the total complex of their geographical experience had evolved in a coherent and consistent manner. The process involved more than a sequence in states of being. It traced a path of becoming. Through a series of generally unconscious incremental adjustments in the modalities of geographical experience, Marie and her peers had progressively reconciled a desire to maintain a familiar life-style with the necessity for innovation in adapting to changing personal and environmental circumstances.

Could this emerging perspective, developed from an intensive experiential study of five individuals, serve as a baseline for a developmental geography of growing old? Exploratory findings clearly require cautious interpretation, particularly when it comes to making more general inferences. Although multiple observations were made, the study involved few participants. The relatively brief three-year span of my work with Marie and her peers renders conclusions regarding developmental change necessarily speculative. Replication is also essential. However, here I am concerned with moving beyond these standard epistemological considerations. It is now several years since I last saw Marie. The vision provided by reflection has served to bring substantive issues I ignored in my original interpretation into clearer focus. Two are elaborated here: the role of contextual factors in uniquely shaping each older person's geographical experience; the thorny problem of differentiating among the effects of aging, membership in a particular age cohort, and 'ageism.'

Contextual Aspects of Geographical Experience

Consider the *personality* context of each older person's life. Marie's effusive outgoing personality dictated a life-style of frenetic action. She toiled from dawn to dusk at a home-based dress repair business, was an active member in clubs and societies, and often extended herself to the point of exhaustion in pursuing a rigorous daily routine. At the same time she was timorous and alienated from the contemporary neighborhood. This was no longer the place she had known. At night the space outside her home became suffused with feelings of threat. She became reluctant to venture forth and would transform her home into a fortress before retiring. After locking all her doors and further securing them with rope, she slept with a loaded pistol readily at hand and a steel ball secured to her wrist by a leather thong. More subtle infusion of personality within her geographical experience was manifest in a strong familism. This was apparent in her benign protective monitoring of the activities of neighborhood children as they passed through the surveillance zone outside her home, and in the environments she chose to inhabit in fantasy. She tended to reconstitute and savor family places, both joyous and tragic; the piazza of 1933 where she danced with her children; the scene of an anniversary celebration; the solemnity of an Arlington graveside when a son was buried – all events and places reinforcing a sense of worth as a loving, successful mother.

Stan, by contrast, was not limited in his wanderings around the neighborhood by any nocturnal timidity. He was more stoic than Marie: A dogged 'work' ethos pervaded his personality and was reflected in the monotonous rhythm of an almost invariant post-retirement routine of daily trips from bar to bar. Even his geographical fantasy typically involved work environments of his past, extending back to vivid images of toiling in his youth behind a team of straining horses in a Polish field. Clearly, variation in personality results in a great diversity of emphasis in the manner in which the modalities of geographical experience find expression in each person's life.

Understanding the infusion of personality within geographical experience entails consideration of the older person's *autobiographical* context. Each life history represents a stream of experience. A residue of rhythms and routines evolved over the lifespan furnishes a template framing a lifeworld in the present (Buttimer, 1976). Here, the sheer length of older people's lives poses a dilemma. To some extent one can generalize about infant behavior. From the breast to the nursery there is much commonality in human experience. As children mature, however, experience diverges – one adolescent remains single and goes to college, another marries and works in a factory. After seventy or eighty years, life paths have become completely individualistic, each defining a unique reservoir of experience. Generalization becomes exceedingly tenuous.

The significance of personality is further elaborated in considering the role of the *physical* context in framing and enriching geographical experience. Clearly, design of the physical context sets limits on the older person's pattern of action. The physical environment also conditions orientation and becomes the repository for a distinctive array of feelings. Beyond this, however, in the process of living within a place, older people gradually develop a reservoir of environmental cues evoking geographical fantasy. Some cues were self-selected. There were Marie's scrapbooks, the clock which Raymond kept on his mantle informing him of the time in Tokyo where his son resided, prominently displayed photographs of family, mementoes of fondly remembered trips, and other treasured artifacts. Other cues were inherent in the neighborhood context. These included important buildings, such as the dilapidated hall where Marie had once danced, and other environmental features of significance within personal history. All these cues were selectively perceived and utilized in sustaining a pattern of geographical experience which both expressed and at the same time reinforced identity and rootedness in place. Indeed, the physical context as subjectively constituted within the participants' experience became an extension of their personalities. Neighborhood transition was ultimately a threat to the self.

When one acknowledges that lives are not only intertwined with place but also with people, the contextual issue becomes even more complicated. Older people live in diverse *family* contexts including two-person households, the traditional multi-generational family, age peer group 'families' of friends, and a variety of other living arrangements (Treas, 1975; Sussman, 1976). Living arrangements and their associated social milieu may profoundly influence geographical experience. Some older people are alone and, by choice or fate, have limited access to social support. Their actions are limited by the absence of rides and other resources a family provides. If there is no one to visit and limited opportunity for vicarious projection into the lives of others, reflective geographical fantasy may become a dominant modality. On the other hand, the geographical experience of a person like Evelyn – the fulcrum of a vast family network providing both direct and indirect support (Osterreich, 1965) – embraces both occasional trips to distant locations where her children reside and a rich world of projective geographical fantasy.

Social involvement extends beyond the family. Most older people's geographical experience reflects their allegiance to a variety of social networks. In particular, geographical experience may be molded by the social mores of a local *community* context. Significant regional variations exist in community conceptions of appropriate behavior for an older person. These are internalized by elderly residents (Lozier, 1975; Lozier and Althouse, 1975). Thus Marie's geographical experience is in part an accommodation to a set of expectations inherent within the culture of the particular French-Canadian neighborhood community of which she is a member. These expectations prescribe an active participation in her church (although absence from services during inclement weather is sanctioned); the development of strong concern for home and neighborhood space; and the maintenance of affinity for the locality in Quebec from which she and many of her neighbors originally migrated.

Finally, the pervasive influence of a specific *societal* context must be acknowledged. This is most apparent when aging is viewed cross-culturally (Cowgill and Holmes, 1972; Gutmann, 1977). The participants' geographical experience is one manifestation of a self-perceived role within a technological society which until recently devalued its nonproductive members. Stan embodied the internalization of an ethos of redundancy: 'Funny world, I tell you that. You struggle, struggle, work, work. Then when you get through work, you're ready to die. That's the end of you.' The resulting alienation encourages physical withdrawal and sanctions introspection and increasing emphasis upon vicarious forms of environmental participation. A societal image becomes a self-fulfilling prophesy.

Aging, Age Cohorts, or 'Ageism'?

The influence of the societal context on the evolving mosaic of the older person's geographical experience introduces a second set of issues which loom more prominently in retrospect. These are an extension of the perennial nature-nurture dilemma. How can one distinguish among changes attributable to inherent biological and psychological processes of aging – changes stemming from membership in a specific generation or cohort of older people, and changes resulting from the impact of the particular societal context in which the person is growing old? This knotty trilemma has been a major focus of debate in gerontology (Neugarten and Datan, 1973; Schaie, 1977).

Certain biological and psychological changes are attributable to aging of the human organism (Finch and Hayflick, 1977; Birren and Schaie, 1977). For example, there is a propensity for collapse of the spinal column, reduced lung capacity, calcification of ligaments, reduced circulatory system capability, and sensory decrements in sight, hearing, taste, touch, and smell. Each of these changes modifies the individual's experience of the physical environment. Impairments, however, are selective. Many older people do not experience them all, nor do transitions occur at comparable rates. For these reasons, it is difficult to specify particular physiological changes as universal attendants of growing old, and to link such changes to consistent transitions in geographical experience.

The problem of physiological variability is compounded by age-cohort differences. The geographical experience of Marie, born in 1890, is different from that which a person born twenty years later can anticipate when she attains 83 years of age. Different cohorts of elderly people experience distinctive life histories in terms of the events which impinge upon them during critical phases in their lives. Marie's generation was profoundly influenced by the experience of raising children during the Great Depression. For people born twenty years later this same event was experienced from the vantage point of late adolescence and hence differently incorporated into autobiography. Recognizing cohort differences in the geographical experience of growing old implies that revelations about today's elderly may be of limited value in understanding those who follow them, even in the same neighborhood.[5]

A complementary aspect of this argument lies in considering the impact of the contemporary societal environment at the time when the individual is growing old. The societal context of aging in 1974 when I spent much time with Marie was different from what it will be in 1994 when a person twenty years her junior will be the same age. For Marie, being 83 years old in an inner-city neighborhood was to accommodate to an ethos of 'ageism' in which her place was defined by an array of societal stereotypes and diverse forms of overt and covert discrimination (Butler, 1975). Her geographical experience with its emphasis upon vicarious environmental participation in displaced milieu may be understood as a creative adaptation to this situation. That many of her peers had made similar accommodations may indicate no more than limited temporal-spatial consistency in the geography of growing old.

Perspective

Looking back on my work with Marie and her peers with the wisdom of hindsight, it is tempting at first to conclude that the insights I obtained are suggestive, even tantalizing, but essentially fragile. To reach this conclusion, however, would be to remain within the limiting confines of a narrowly defined empiricist view and to ignore the humanistic emphasis of the study. Descriptive understanding of Marie's involvement within the spaces and places of her life provides an authentic representation of the reality of growing old as experienced by one person dwelling within a specific spatio-temporal setting. Much of what is significant in her geographical experience stems from her unique life experience and the kind of person she has come to be. It is beyond the realm of generalization. The essence here lies in *understanding.* Understanding, it can be said, is the deeper level of awareness which arises from drawing close enough to a person to become a sympathetic participant within her life-world and to have her integrally involved in one's own. Understanding has value in itself: indeed, it should be a major goal of scholarly endeavor.

Understanding individual experience in all its complexity is a necessary prelude to constructing sensitive theory. This is a major theme I

, 1975b. 'Space and Place: Humanistic Perspective.' In C. Board, R.J. Chorley, P. Haggett and D.R. Stoddart (eds.), *Progress in Geography*, Vol. 6, pp. 212-52. London: Edward Arnold.
1976. 'Humanistic Geography,' *Annals of the Association of American Geographers*, 66, 266-76.
1977. *Space and Place: the Perspective of Experience*. Minneapolis: University f Minnesota Press.
Eckartsberg, R. 1971. 'On Experiential Methodology.' In A. Georgi, W.F. ischer, and R. Von Eckartsberg (eds.), *Duquesne Studies in Phenomenological ychology*, Vol. 1, pp. 66-79. Pittsburgh: Duquesne University Press.

have sought to illustrate in this essay. The quest for understanding, through its very process – the revelation of essential themes through interpersonal dialogue – provides for internal critique in the process of data generation. Potential for the naive stereotyping which characterized my initial view of Marie's geographical experience is reduced.

The process of abstraction from the taken-for-granted coherence of direct experience to a formal conceptualization involves a qualitative transition in both the language and substance of knowing. Translation from the prereflective understanding of everyday life to the language of social science mirrors an operational categorization of what is in experience an undifferentiated coherent whole. This process reduces experience, but it is valuable for comprehending and communicating the complexity of lives if the categories represent authentic, inherently experiential themes. However, to the degree that a conceptualization becomes the imposed structuring of an outsider who is separated from the experience, an inevitable discordance arises between the two levels of knowing. Moreover, over time, as caveats and elaborations such as I have presented are incorporated, what originates as an authentic translation from experience may come to assume a life of its own, independent of the existential reality from which it originated. By manipulating emergent themes in a search for more sophisticated *explanation,* there is an ever-present danger of falling into the reductionist trap of considering as legitimate only those aspects of experience which can be generalized. In so doing much of the richness of individual experience is cast aside. An understanding of person as subject – the author of a unique biography lived within a colorful lifeworld – is rejected. Ironically, as we progress toward more sophisticated explanation in developing a geography of growing old, our understanding may become progressively impoverished. Yet to the humanist geographer it is the understanding that is ultimately most important.

Notes

1. In this essay an older person is defined as a person over 65 years of age. This designation is an arbitrary choice. Chronological age is, of course, only one of many possible measures of aging. There are important differences among biological, psychological and social aging processes. Moreover, rates of aging in these domains vary considerably among individuals.
2. These vignettes provide a primary evidential basis for a study such as this, for they reveal the 'life history' of the experience (Becker, 1958; Rowles, 1978b). Unfortunately, space precludes the presentation of all this material here.
3. Confirmation of this pattern in ongoing research I am conducting with a

population of elderly residents of an Appalachian community suggests that increased propensity for long-distance trips may be a characteristic feature of the older person's experience, at least in the United States, particularly in the years immediately following retirement.

4. Occasional trips constituted a significant exception to this overall pattern. However, anticipation of these trips often generated vivid projective geographical fantasy, and fond recollection of such vacations was an important source of reflective geographical fantasy.

5. This problem raises sobering methodological questions regarding inferences of developmental change derived from cross-sectional rather than longitudinal studies (Schaie, 1967; Maddox and Wiley, 1976).

References

Becker, H.S. 1958. 'Problems of Inference and Proof in Participant Observation,' *American Sociological Review, 23,* 652-60.

Birren, J.E. and Schaie, K.W. 1977. *Handbook of the Psychology of Aging.* New York: Van Nostrand Reinhold.

Bollnow, O. 1967. 'Lived-Space.' In N. Lawrence and D. O'Connor (eds.), *Readings in Existential Phenomenology,* pp. 178-86. Englewood Cliffs, NJ: Prentice-Hall.

Butler, R.N. 1963. 'The Life Review: An Interpretation of Reminiscence in the Aged,' *Psychiatry, 26,* 65-76.

—, 1975. *Why Survive? Being Old in America.* New York: Harper & Row.

Buttimer, A. 1976. 'Grasping the Dynamism of Lifeworld,' *Annals of the Association of American Geographers, 66,* 277-92.

Coleman, P.G. 1974. 'Measuring Reminiscence Characteristics from Conversation as Adaptive Features of Old Age,' *Aging and Human Development, 5,* 281-94.

Cowgill, D.O. and Holmes, L.D. 1972. *Aging and Modernization.* New York: Appleton-Century-Crofts.

Downs, R. and Stea, D. 1973. *Image and Environment.* Chicago: Aldine.

—, 1977. *Maps in Minds: Reflections on Cognitive Mapping.* New York: Harper & Row.

Entrikin, J.N. 1976. 'Contemporary Humanism in Geography,' *Annals of the Association of American Geographers, 66,* 615-32.

Finch, C.E. and Hayflick, L. 1977. *Handbook of the Biology of Aging.* New York: Van Nostrand Reinhold.

Godkin, M. 1977. 'Space, Time, and Place in the Human Experience of Stress,' unpublished PhD dissertation, Clark University, Worcester, Massachusetts.

Gutmann, D. 1977. 'The Cross Cultural Perspective: Notes Toward a Comparative Psychology of Aging.' In J.E. Birren and K.W. Schaie (eds.), *Handbook of the Psychology of Aging,* pp. 302-26. New York: Van Nostrand Reinhold.

Hochschild, A.R. 1973. *The Unexpected Community.* Englewood Cliffs, NJ: Prentice-Hall,

Howard, I.P. and Templeton, W.B. 1966. *Human Spatial Orientation.* New York: Wiley.

Lewis, C.N. 1971. 'Reminiscing and Self Concept in Old Age,' *Journal of Gerontology, 26,* 240-3.

Ley, D. 1977. 'Social Geography and the Taken-for-Granted World,' *Transactions of the Institute of British Geographers, 2,* 498-512.

Ley, D., and Samuels, M. (eds.). 1978. *Humanistic Geography: Prospects and Problems.* Chicago: Maaroufa Press.

Lozier, J. 1975. 'Accommodating Old People in Society: Examples and New Orleans.' In N. Datan and L.H. Ginsberg (eds.), *Lifespa mental Psychology: Normative Life Crises,* pp. 287-97. New Yo Press.

Lozier, J. and Althouse, R. 1975. 'Retirement to the Porch in Ru *Aging and Human Development, 6,* 7-15.

Maddox, G.L. and Wiley, J. 1976. 'Scope, Concept and Methods Aging.' In R. Binstock and E. Shanas (eds.), *Handbook of A Sciences,* pp. 3-34. New York: Van Nostrand Reinhold.

McMahon, A.W. and Rhudick, P.J. 1967. 'Reminiscing in the A tional Response.' In S. Levin and R.J. Kahana, *Psychodyn Aging: Creativity, Reminiscence, and Dying.* New York: I University Press.

Montgomery, J.E. 1977. 'The Housing Patterns of Older Peo (ed.), *The Later Years: Social Applications of Gerontolo* Cole.

Moore, G. and Golledge, R. (eds.), 1976. *Environmental K* Pa.: Dowden, Hutchinson, and Ross.

Neugarten, B. and Associates. 1964. *Personality in Middle* Atherton.

Neugarten, B.L. and Datan, N. 1973. 'Sociological Persp In P. Baltes and K.W. Schaie (eds.), *Life-Span Develo Personality and Socialization,* pp. 53-69. New York:

Osterreich, H. 1965. 'Geographical Mobility and Kinsh R. Piddington, *Kinship and Geographical Mobility.*

Pastalan, L. 1971. 'How the Elderly Negotiate Their E at Environment for the Aged: A Working Confere Utilization and Environmental Policy. San Juan,

Relph, E. 1976. *Place and Placelessness.* London: Pi

Rowles, G.D. 1978a. *Prisoners of Space? Exploring Older People.* Boulder, Colorado: Westview Pre

—, 1978b. 'Reflections on Experiential Fieldwork (eds.), *Humanistic Geography: Prospects and P* Maaroufa Press.

—, 1979. 'The Last New Home: Facilitating the Institutional Space.' In Stephen Golant (ed.) *Elderly Population.* Washington, DC: V.H. W

Schaie, K.W. 1967. 'Age Changes and Age Diffe 128-32.

—, 1977. 'Quasi-Experimental Research Desi J.E. Birren and K.W. Schaie (eds.), *Handb* pp. 39-58. New York: Van Nostrand Reir

Seamon, D. 1979. *A Geography of the Lifew* London: Croom Helm.

Shephard, R.J. 1978. *Physical Activity and* Publishers.

Sussman, M.B. 1976. 'The Family Life of Shanas (eds.), *Handbook of Aging and* York: Van Nostrand Reinhold.

Treas, J. 1975. 'Aging and the Family.' I *Aging: Scientific Perspectives and Sc* Nostrand.

Tuan, Y. 1975a. 'Images and Mental M *Geographers, 65,* 205-13.

3 IDENTITY AND PLACE: CLINICAL APPLICATIONS BASED ON NOTIONS OF ROOTEDNESS AND UPROOTEDNESS

Michael A. Godkin

The places in a person's world are more than entities which provide the physical stage for life's drama. Some are profound centers of meanings and symbols of experience. As such, they lie at the core of human existence. This essay, drawing upon the life experiences of several alcoholics, demonstrates ways in which places become reservoirs of significant life experiences lying at the center of a person's identity and sense of psychological well-being.

As a context for the essay, some of the relevant literature examining the significance of places in the life experiences of people in general is first reviewed. Second, accounts of the lives of three alcoholics are drawn upon to demonstrate the ways in which sensory images of places become woven into the life experiences and identities of individuals. Third, the possibility of developing a modality of psychotherapy based on place-image chronologies is discussed. Finally, some implications of the findings are examined with respect to issues of the design and planning of therapeutic settings and the possible contributions of behavioral geography to an understanding of human behavior and experience.

The Human Experience of Place

The existential significance of places has been acknowledged in various definitions which characterize places as the 'focus of meanings or intention, either culturally or individually defined' (Relph, 1976, 55), or entities which 'incarnate the experience and aspirations of people' (Tuan, 1971, 281). This essay defines place as a discrete, temporally and perceptually bounded unit of psychologically meaningful material space.

Much of the literature on the experiential dimensions of place has focused on those places to which shared meanings or common symbols are attached by certain groups of individuals. These can be places which evoke some sense of belonging to a social group and provide a sense of group identity. Such places exist at various scales. Relph (1976) has identified, for example, the Red Square in Moscow, Niagara Falls and the Acropolis as national symbols of common experience which foster a sense of national unity and pride. In some countries, at a more local level, the corner grocery or soda spa exist as important symbols of neighborhood and community identity. Other researchers have alluded to the existence of universal, shared place meanings which transcend political or social group identities. Bachelard (1964), for example, postulates that real or imagined places, characterized by the four primary substances of fire, air, water, and earth, are one foundation of human experiences and actions. Images of mountains, for example, evoke feelings of power, while images of fire indicate a feeling of paralysis.

As well as embodying shared experiences, places may hold ascribed meanings unique to individuals. Relph (1976, 37), for example, has suggested that 'there may indeed be no common knowledge of them [places] ; rather, they are defined by special and particular significances for us and can be remembered rather than immediately present.'

In fact, the importance of attachments to place has been acknowledged in the psychiatric literature as significant in the development of self-identity (e.g. Searles, 1960; Wenkart, 1961). Little, if anything, however, has been written about the mechanisms and processes by which images of place become integrally woven into an individual's sense of self. Drawing on autobiographical accounts of three alcoholics clinically diagnosed as psychologically stressed, this essay examines the role of place in people's development of self-identity and feeling of well-being.[1]

The Role of Places in the Lived Experience of Alcoholics

Recovering alcoholics were chosen as the focus for this study.[2] The choice was based on the possibility of demonstrating the importance of images of place in both the initial development of a negative self-image and accompanying alcoholic behavior and subsequent recovery process and gradual emergence of a positive sense of self-identity. Because of the in-depth nature of the interviews, the study was restricted to three participants.[3] This limited number obviously restricts generalizations about the exact nature of places which are associated with either stressful experiences or a sense of well-being. Tentative statements, however, can be made.

Specifically, the participants in this study were Doris, 43 years old; Ellen, 63; and Ben, in his mid-fifties. For the purpose of illustration, the majority of quotations in this essay are drawn from the accounts of Doris. The relationship of places to the development of psychological well-being on the one hand, or psychological stress on the other, can be subsumed under two key notions that arose – *uprootedness* and *rootedness.*

Uprootedness for the Alcoholic

The alcoholic's day-to-day life is often dominated by a lack of self-worth. He or she may feel lost and question his or her identity and value as a person. In part, this frequent confusion of self is characterized by *uprootedness* – a sense of non-belonging to place. Places in this mode of experience are perceived as threatening; they interfere with the integrity of one's identity.

Consider, for example, Doris, whose mother was confined to a mental institution when Doris was four. Doris was placed in an orphanage and lived there for the next ten years. She vividly remembers the orphanage as a place: 'The floors were hardwood – huge rooms very bare, very cold, not the closeness . . . big dorms. The rooms were big, big, but you were so small and lonely . . . Sometimes I wouldn't let it get to me.' Doris did not feel she belonged in the orphanage. Rather, she experienced it as a threatening, unsafe place. She was small – the rooms large. 'Sometimes,' she explains, 'you'd lay in bed at night, you'd be so lonesome, and you'd cry yourself to sleep.' Doris was detached from the orphanage and isolated from place. The extent of her uprootedness is depicted by a sudden recollection of the orphanage's halls which trigger the memory of a visit from her father:

> Now something is coming through my mind . . . the hallways, the long hallways . . . the wood. Christ! I haven't thought of this for a long time. I remember my father coming up when I was really young there, crying like hell because I didn't want to stay and he was crying and then I didn't see him for a long time because I could imagine it hurt when he saw us there. He really didn't want to put us there, but he did.

Though Doris left the orphanage at the age of fifteen, the unpleasant feelings associated with her stay there frequently permeated her adult life. The places where she lived after the orphanage often evoked similar descriptive responses: frequently these places were cold and she felt isolated, unwanted, and unimportant. Describing her present home, Doris explained that it was mediocre and projected no warmth. It seemed not to possess the elements of at-homeness that a house normally has:

> Part of my house reminds me of the orphanage – it's blah. I guess there's no love when I think of the rooms and the plainness. This is a cold house. With the orphanage it was always cold – the hardwood floors – not homey.

In a similar way, non-residential places in Doris's present life sometimes reminded her of the orphanage. For example, the mental institution where she had researched her mother's medical records reminded Doris of the orphanage interior and her frequent experiences of depression. This hospital, she explained, 'was dead, the floors were dark, its decor dreary and depressing.' Similarly, Doris had felt uncomfortable attending classes at the local community college but could not precisely determine the reason. When describing the orphanage hallways, Doris suddenly drew a connection between these halls and the interior of the college: 'I didn't place the [uncomfortableness] then, but it was a feeling like the orphanage – the hallways, going up the stairs.'

Ellen and Ben, the two other alcoholics interviewed in the study, also expressed frequently a sense of uprootedness in their comments on place. Ben, his childhood characterized by a sense of not belonging, said that one foster home in which he was placed 'offered a minimum of human care . . . and a maximum of criticism, hollering and screaming.' Ellen, unwanted by her mother, summarized her sense of alienation in her description of the apartment where she and her mother lived. 'I was miserable there,' she explained, 'it was full of cockroaches, bedbugs, and rats.' In contrast, she described a foster home where she lived for a time: '[It was] a nice country home where they showed affection. It was full of love.'

In part, then, the psychologically stressed alcoholic feels alienated because some life experiences may have involved living in places where he or she felt unwanted and to which there was no sense of belonging. As with Doris, a sense of being apart and different from one's surroundings can trigger self-doubts and feelings that 'there is something wrong with me.' Images of such places which were perceived as in some way threatening to self-identity become associated upon reflection with a sense of uprootedness – of not having a place where it is possible to 'anchor' one's uniqueness.

At particular times such images may come to conscious attention and generate a sense of worthlessness and associated feelings of anxiety, fear, shame, sadness, or anger. A self-reinforcing stress spiral develops. Each negative experience reinforces an emerging sense of worthlessness and increases the likelihood that certain places in the future will be seen as symbolic reminders of past places associated with threatening experiences. Such was the case with Doris, whose present home and the college which she presently attends were symbols of and evoked similar experiences to the orphanage. Increasingly, places in the potential alcoholic's present life become reminders of past places associated with negative experiences. Stressful emotions become more intense and persistent. Eventually, a point is reached when the potential alcoholic selects excessive alcohol as a means of coping with the discomfort of his or her feelings.

Rootedness for the Alcoholic

Although the alcoholic generally feels alienated and without personal worth, there is sometimes a positive sense of self, usually lost to immediate memory. In part, this feeling of worth is characterized by an underlying sense of *rootedness* – a feeling of belonging to place.

Rootedness, interviews suggest, is less common in the alcoholic's experience than uprootedness. Still rootedness is present, especially in relation to places which the alcoholic knew or knows as *refuges* – safe places in which the person escapes an otherwise unhappy existence. In Doris' childhood, for example, this place was associated with the land behind the orphanage where the children would be taken for picnics:

> I always loved picnics. The picnics to me meant being free and being out, especially in the spring when the violets bloomed. You see, I always loved the flowers. I remember the brook when we went on picnics. We had to go through a pig farm and the cows and it stunk like hell. I remember slipping one day – yuk! But once you got beyond that you were in the woods.

On the picnics Doris experienced a stronger sense of harmony and at-homeness with her surroundings than at the orphanage itself. There was less the separation between her and place than she experienced in the orphanage building proper. In the orphanage Doris was one small cog in a large impersonal institution. The picnic experience, in contrast, fostered a sense of freedom, and wish to open outwardly to the environment and events at hand:

> You were more free on the picnics; there wasn't so much discipline. You still had to do what you were told, but I could go off by myself near the brook and I remember the violets and it fascinated the hell out of me. Skunk cabbage – I just saw it now. It flashed through my mind – and the trees. The skunk cabbage stunk but I liked it – being out with the sky and the ground.

After leaving the orphanage, Doris spent two years living in a small town with her father and the family of his new woman friend. Overall, a sense of uprootedness pervaded her experience there. At the same time, however, there were two places where she felt some degree of rootedness – the cellar of her home and the nearby railroad tracks – both seemingly unusual places of attachment, but significant in Doris' experience. 'I felt safe downstairs,' she said, describing the cellar, 'and there was a door right from the cellar and you could go outside . . . The cellar was closed, keeping me secure.' The railroad tracks offered Doris rootedness for another reason: they provided a private place of freedom and links with the world of possibilities beyond her own limited experience:

> I liked taking walks down by the railroad tracks. It was like being free and, of course, when you went with kids it was like playing games. If a train was coming we'd wait and then run like hell or sit on the fence near the cows, waiting for the train to go by. To me that was great and you could wave to the conductor. There was a feeling, maybe, of wanting to go someplace, I don't know – of being on the train and being free and going someplace.

Places of rootedness were also described by Ellen and Ben. Ellen vividly recalled the basement workshop of her foster father, whom she grew to love because he showed her much affection. 'Downstairs,' she explained, 'was my whole world.' The best part of her day, she explained, began when her foster father returned home and they went down to the workshop 'to hammer nails and get dirty.' Ben had clear memories of the last place his family had lived in before his mother died. Particularly, he remembered the space in front of the house where he helped his father clear stones. 'I guess,' he explained, 'it was kind of like a safe, good place to be.' Ben also remembered another place near the house where there was 'a brook and rocks and trees.' Like Doris' railroad tracks, this place provided both excitement and a place to be one's self:

> It was a place to explore and, I guess, sometimes a place to hide, a

> place to retreat to if you wanted to avoid authority, parental figures, grown-ups. I guess you'd have to say it was a private kind of place because you could be whatever you were. I have a picture in my mind of sitting on a rock with high, white sneakers and blond bangs and a cat on my lap and petting the cat and the picture says I was very pleased with myself.

These illustrations of rootedness for the alcoholic, as few as they are, indicate that positive images of place can provide a concrete focus for the attachment, retention and development of self-identity. Experiences supporting a sense and coherence of self are 'captured' and retained in memory partially as an image of the place where the positive experiences originally occurred. In other words, a sense of self-identity is partially incorporated into an individual's being when anchored to the place in which it was experienced. Unfortunately for the alcoholic, places associated with rootedness are few in number and often may be lost to immediate memory. A therapy grounded in rootedness and uprootedness, therefore, attempts to strengthen positive images of place and defuse negative recollections.

Implications for Therapy

The apparent importance of rootedness and uprootedness in the lives of alcoholics indicates certain possibilities for the use of the notion of place as a basis of a therapeutic modality for the treatment of their psychological stress. Findings from a pilot study support this premise.[4] The therapeutic technique involved patients recalling and describing in chronological sequence significant places in their lives – e.g. orphanages, foster homes, homes of family, friends, relatives, schools, and work places. Patients were told to close their eyes and share immediate thoughts. For example, consider the following questions concerning the first place in which the patient could remember living:

> I want you to cast your mind back and think about the first place in which you remember living. Look at it from the outside and describe it to me (cues can be given where necessary, referring to the shape, size and color of the building).
>
> What sort of feeling do you have as you describe the house?
>
> Now take me inside your house and as we open the door describe to me what you see.

Let us go into a room where you liked to spend time and describe it to me. Do you remember any particular sounds or smells?

Were there any rooms which you did not particularly like. Could you describe them?

Did you have your own bedroom? Describe the room where you used to sleep. Did you like being there?

Now, could you describe some of the other rooms in the house? (if not described already, mention the kitchen, den, living-room, parents' and siblings' bedrooms, basement, and elicit images of all senses).

Did you spend much time in your house?

Now, let us go outside. Describe the surroundings to your house. Did you play near your house? Were you usually alone or with friends? Did you spend much time there? Did you enjoy playing near your home? Why do you think that was the case?

Were there any other places where you used to go to play? What did they look like?

Was there any particular place to which you would go when you were upset? Could you describe it to me? Did you prefer to play outside or in your house? Why do you think this was the case?

It appears that clinical interviews focused around such questions allow a client to rediscover and re-examine a forgotten sense of coherence in personal identity which was originally a part of experience in places associated with positive memories. Such places are those in which self-identity was originally anchored and to which an individual feels a sense of belonging or rootedness. The chronological recall of perceptual images of places associated with significant positive experiences helps a patient 'rethread' his or her own identity, which has been forgotten and fragmented in a few isolated and dispersed places. In this way, continuity in and inner sense of belonging to place, paralleling continuity in self-identity, can be established.

The recall of places associated with negative memories, on the other hand, facilitates a process of catharsis. A greater degree of affect response appears to accompany conversations based on places rather than people. Defense mechanisms which normally inhibit expression of negative feelings are often circumvented by basing interviews on places. The smaller degree of perceived threat in such interviews facilitates the expression of suppressed, stressful emotions. In this way a patient is able to become free and cleanse himself of emotions which before were difficult to share.[5]

The present author found an enhanced ability on the part of participants to recall both positive and negative experiences relating to places (Godkin, 1977). Possibly, this result is the outcome of emphasizing concrete, tangible descriptions rather than abstractions. Of particular importance is the ability to recall forgotten place experiences which reflect the existence of a *coherent* sense of self. For example, Doris' immediate memory of her days in the orphanage was dominated by images reflecting its disciplined life-style. It was only when she was asked to describe the outside grounds that she began to remember the happy times she spent on picnics. Doris referred to herself as a 'loner' in early interviews: she said she had no friends and that nobody liked her. But during a ride to some places in which she had once lived, she suddenly pointed to a house which triggered memories of days she had spent with other girls who had lived there. 'Gee,' she exclaimed, 'a lot of memories are coming back now! When I think of it, I used to have a lot of girl friends.' Doris had remembered a part of her life which she had previously forgotten – a part of herself which she obviously liked. She was recalling herself as an outgoing child who was liked by others and had friendships.

The potential application of the recall of places as a therapeutic basis is perhaps most strikingly demonstrated by Ellen. During the course of the interviews Ellen became able to recall happy memories from her married life which had become clouded by the unhappiness of excessive drinking, divorce, and illness, and thereby lost to immediate memory. 'I think', she said, 'really and truly, talking to you like this has made me see how I was all clammed up.' Then she began to recall a place near the ocean where she and her ex-husband used to own a summer home:

> I didn't realize how much fun I was having when I would creep out at four in the morning. I always had a can of worms on the porch. I'd throw a line out – the water was still. You could see the eels and frogs and beautiful birds. I liked to walk in the dunes – so many things I'd forgotten. Making love on the beach was a great thing when I was young. All of those things – you can't go back, but you have those memories locked up tight. It's over but it isn't really – your memories. You don't just become a nothing. You have feelings which are much more meaningful. So, when you get in my position and you think there may not be a whole lot more of them, instead of being full of self-pity, you remember them with joy, with a feeling that just

> engulfs me from my toes to my head. It's just a feeling that for one fleeting moment it makes life all worthwhile.

Ellen's recall indicates the gradual rediscovery of parts of herself which reflect a person she likes and allow her to feel as though her life has meaning – it is 'worthwhile.' Being able to share the unhappiness and, most importantly, rediscover the happy times, had allowed her to 'see myself much better, without the hurts – balancing.'

Implications for the Design of Therapeutic Settings

The notions of uprootedness and rootedness have had value in redesigning an existing hospital setting and creation of a Palliative Care Unit for terminally ill cancer patients.[6] A major focus of palliative care is the alleviation of pain and discomfort from physical symptoms. Just as important, however, are the emotional and spiritual needs of patients and their families. A family therapist, thanatologist, social worker, and chaplain are each integrally involved in attempts to enhance patient and relatives' psychological well-being. Paralleling this attention to the human needs of people is the aim to create a homelike environment reflecting the care and attention paid to the well-being of patients.

An assumption was made that the stressful feelings of fear and isolation which many dying patients experience are exacerbated by traditional care facilities which, in essence, remove a person physically from the familiarity of his regular surroundings. Such an approach reinforces the notion that dying is the antithesis of living – that it is a totally separate and stressful experience, the constituents of which are unrelated to the rest of one's life. Instead, the aim is to have the patient see dying as a difficult but inescapable stage in life's progression. As such, the reduction of fear and isolation and the enhancement of well-being should be fostered by establishing a sense of continuity in the dying person's life. An emphasis on continuity can be reflected in and symbolized by the construction of a physical setting which incorporates items which are familiar to patients.

The components of a homelike environment in the unit include a 'family kitchen' where family members can prepare the patient's meals, which are then eaten in the 'dining-room.' Patients spend time together in the 'den,' where on Fridays they can gather for alcoholic beverages and an appropriately named 'happy hour.' There is a pleasantly furnished 'family room' where family members can spend time with others facing similar situations. Two 'sleep-over' rooms, containing double beds, provide an opportunity for spouses to sleep together as they might at home. Children are encouraged to visit their parents and grandparents, and a 'children's room' is furnished with equipment which facilitates expression of their experiences – e.g. telephones, puppets, paper, and paints. The sound of children's voices appears to be of particular importance for the dying patient. They remind him of his position in life's cycle. He gains some comfort from the knowledge that life itself continues even if his own must end.

Patient and family's spiritual needs – which often have been particularly important components of their lives prior to the illness – can still be expressed during the weekly services in the unit's 'chapel.' Two small alcoves (which are used as a stretcher area and nursing sub-station on the other floors of the hospital) have been converted to sitting areas containing plants, aquariums, paintings, and other symbols of life. Physical barriers which were initially designed to delineate professional space from patient space have been removed. The implicit message communicated by the new design is that patients and their families can, at this time of crisis, enlarge their familial boundaries and include the staff of the unit.

In addition, to foster a sense of purposefulness, destination places have been designated which encourage patients to leave their rooms. For example, patients are encouraged and aided when necessary to use a 'family room' when they receive visitors. The room is an opposite wing to the main patient corridor and necessitates a degree of effort to get there. Patients are also encouraged to use the hospital cafeteria and gift shop.[7] As much as possible, they are encouraged to go outside in the hospital grounds or even return home, if the necessary family and medical supports exist.

In part, as the above description indicates, the design of the Palliative Care Service was organized around the notions of rootedness and uprootedness. The design reflects a tacit assumption that physical surroundings are an integral component in the experiences and senses of well-being of patients and their families. In particular, the homelike environment provides a tangible and immediate sensory image which suggests that patients and families are in a safe place of *rootedness* where they feel cared for and secure. The construction of surroundings which are symbolically familiar fosters a sense of continuity in the life experiences of patients and families. These surroundings provide a link to the more hopeful and happier stages of the patient's life, prior to the onset of illness. Such continuity helps patients and family members retain a stable and coherent sense of self at a time of vulnerability when the uncertainty and instability of the unknown future could fracture a fragile sense of identity.

Implications for Environmental Theory

This chapter has attempted to outline the nature and implications of certain relationships between places and human experiences. In particular, it has been suggested that an examination of the existential dimensions of place has certain practical value: facilitation of more comprehensive understanding of experiences which are psychologically stressful; developments of therapeutic intervention techniques based on place-image chronologies; the possibility of designing human environments which support an individual's sense of well-being.

Only recently have behavioral geography and environmental psychology become interested in the experiential dimensions of place. This essay, taking discussion one step further, has examined the ways in which places become integral components of an individual's emerging sense of self. The apparent significance of places in the learning and experiences of people reinforces a contention that behavioral geographers can make a significant contribution to the development of theory in the areas of human behavior and experience. Also, the increasing recognition of transactional perspectives in psychology has added weight to arguments which advocate that human behavior and experience can only be understood by examining the interdependencies between an individual and his or her environment. Further, it would seem likely that as behavioral geographers and environmental psychologists begin to understand the nature and variations in place experiences for different groups of people, they can become integrally involved in the planning of human environments which reinforce experiences fostering a sense of well-being.

Notes

1. Psychological stress is considered to be the experience of a psychological disturbance and contending negative emotion when an individual perceives a real or imagined threat in his or her environment. The importance of psychological stress in the development of alcoholism has received wide recognition in the literature (e.g. Shaefer, 1971; Pratt, 1972; Wilsnack, 1973; Ludwig, 1975).

2. It is generally accepted that alcoholism can be a state which endures throughout a lifetime. Sobriety is recognized as only one stage in a continuous recovery process.

3. A detailed discussion of these three individuals and the research procedure is provided in Godkin, 1977.

4. See Deering *et al.*, 1977. This study demonstrates that after place-centered interviews with a psychiatrist and other clinically trained professionals, approximately three-quarters of a sample of 58 patients with drug, alcohol, or phobic problems had shown improvement in function and decrease in symptoms in one or more areas.

5. Said one psychiatrist-collaborator using this approach, 'I had never tried this technique before and there's a lot of feeling that goes with it. It allows one to talk freely without constraint. It bypasses a lot – that is to say, normal defense mechanisms that are involved when talking about interpersonal relationships are bypassed by talking about the geography of place' (from a tape-recorded conversation with George Deering, MD, July 1976, and reproduced with his permission).

6. The Palliative Care Service at the University of Massachusetts Medical Center, Worcester, Massachusetts, is a service for terminally ill cancer patients based on a 'hospice' principle. The emphasis is on *human care* rather than *technological cure*. Aggressive treatments are performed only if they are considered likely to alleviate discomforting symptoms and enhance the quality of the patient's remaining life.

7. One day, a female patient was buying some chocolates in the gift shop and remarked with a satisfied smile, 'This is the first money I've spent in over three months.' In traditional care facilities, a patient is removed from even the most routine daily activities, like spending money.

References

Bachelard, G. 1964. *The Poetics of Space.* Boston: Beacon Press.

Deering, G., Godkin, M., and Mason, E. 1977. 'Geographic Concepts in Alcoholism,' paper presented at the sixth World Congress of Psychiatry, Honolulu, Hawaii, 2 September.

Godkin, M. 1977. 'Space, Time and Place in the Human Experience of Stress,' PhD dissertation, Clark University.

Ludwig, A.M. 1975. 'The Irresistible Urge and the Unquenchable Thirst for Alcohol,' *Proceedings of the Fourth Annual Alcoholism Conference of the National Institute on Alcohol Abuse and Alcoholism,* pp. 3-23.

Pratt, B. 1972. 'Studies Reveal Alcoholism Differs in Males and Females,' American Psychological Association, *Monitor, 3,* 1-2.

Relph, E. 1976. *Place and Placelessness.* London: Pion.

Schaefer, H.H. 1971. 'Accepted Theories Disproven,' *Science News, 99,* 182.

Searles, H.F. 1960. *The Nonhuman Environment.* New York: International University Press.

Tuan, Yi-Fu. 1971. 'Geography, Phenomenology and the Study of Human Nature,' *The Canadian Geographer, 15,* 181-92.

Wenkart, A. 1961. 'Regaining Identity Through Relatedness,' *American Journal of Psychoanalysis, 21,* 227-33.

Wilsnack, S.C. 1973. 'Femininity by the Bottle,' *Psychology Today, 82,* 39-43.

4 THE INTEGRATION OF COMMUNITY AND ENVIRONMENT: ANARCHIST DECENTRALISM IN RURAL SPAIN, 1936-39*

Myrna Margulies Breitbart

> Decentralism is a kind of social organization; it does not involve geographical isolation, but a particular sociological use of geography – Paul Goodman (1965).

On 18 July 1936 Generalissimo Franco attacked the newly elected Spanish Republic and initiated a long civil war which claimed over one million lives. The political and military struggle which was fought on one side by the army, the Church and the wealthy, and on the other by a loose coalition of the middle class and forces on the left, is well documented in political texts (Thomas, 1961; Carr, 1971; Bolloten, 1961). The anarchist social revolution which accompanied this military struggle has been virtually ignored.

Anarchists who initiated a social revolution in July 1936 raised fundamental questions about economic, social, and political freedom. They did not resist Franco 'in the name of "democracy" and the status quo' (Orwell, 1938, 48), nor did they engage in a civil war to achieve slightly improved conditions within a reformed capitalist system. Years of dissatisfaction under the authority of a centralized political regime created an environment eminently receptive to anarchist aims for communal self-managed production, political decentralization, and cultural autonomy.

Within hours of Franco's attack, anarchist peasants and workers seized direct control over rural land, cities, factories, and social service and transportation networks. In many parts of Republican Spain, whole communities and workplaces were largely self-managed by those living or working within them. Approximately two thousand collectives were voluntarily formed in rural Spain, and over 15 million acres of land expropriated between July 1936 and January 1938. Collectivization encompassed more than one-half of the total land area of Republican Spain, directly or indirectly affecting the lives of between 7 and 8

*Portions of this article also appear in Myrna Margulies Breitbart, 'Spanish Anarchism: an Introductory Essay' and 'Anarchist Decentralism in Rural Spain, 1936-1939,' *Antipode, 10-11* (1979), 60-70, 83-98.

million people (Broué and Témime, 1970, 159; Leval, 1971, 80). (See Figure 4.1.)

The social revolution was not, however, confined to rural Spain. For one to three years, local citizen-worker groups acting without governmental or managerial authority transformed large cities like Barcelona into federations of neighborhoods, and factories into communities of work. In Catalonia, various forms of workers' control were extended to industry, rail lines, transportation systems, entertainment facilities, health and service sectors, gas and electric companies, newspapers, industrial suppliers, and retailers. Outside Catalonia, large industries in urban Levante, Asturias, Aragon, and Castile were also self-managed by anarcho-syndicates and socialist unions (Bolloten, 1961, 48-54; Brademas, 1955).

Anarchist decentralism as applied in Spain during these years proved to be an integrated revolutionary philosophy – an entirely new mode of existence – requiring massive alterations in most aspects of urban and rural life. This chapter describes the changes brought about in the *genre de vie* of rural Spain as a result of the implementation of anarchist forms of economic and social organization. Collectivization created new social relations of production, upsetting traditional land ownership patterns, divisions of labor and managerial hierarchies. These changes and the attachment of Spanish peasants to their *patria chicas* (local home areas) affected the use of technology and resources. Decentralist modes of social and technical organization generated vastly different spatial patterns at both the local and regional level.

The effectiveness with which Spanish anarchists self-managed their work and community environments during a period of enormous disruption suggests a capacity for organized social action that has rarely been attributed to the anarchist tradition. The alternative socioeconomic and spatial forms which were implemented in Spain over forty years ago thus raise important questions of contemporary relevance regarding the revolutionary process, the inefficiencies and inequities resulting from centralization, and the potentialities of self-management (worker and community control). The answers which Spanish peasants and workers furnished to these questions compose the most extensive attempt yet made to implement anarchism in a modern context, and may well provide a base from which one can explore contemporary applications of anarchist thought.

Figure 4.1: Major Areas of Collectivization, 1936

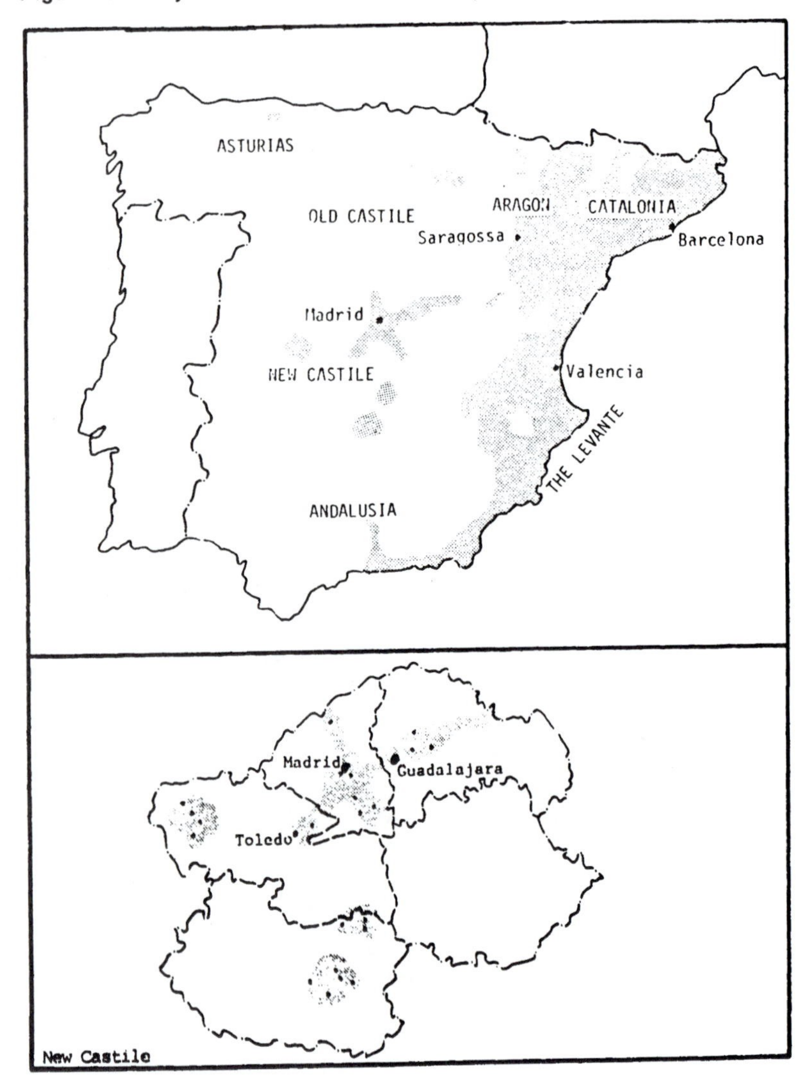

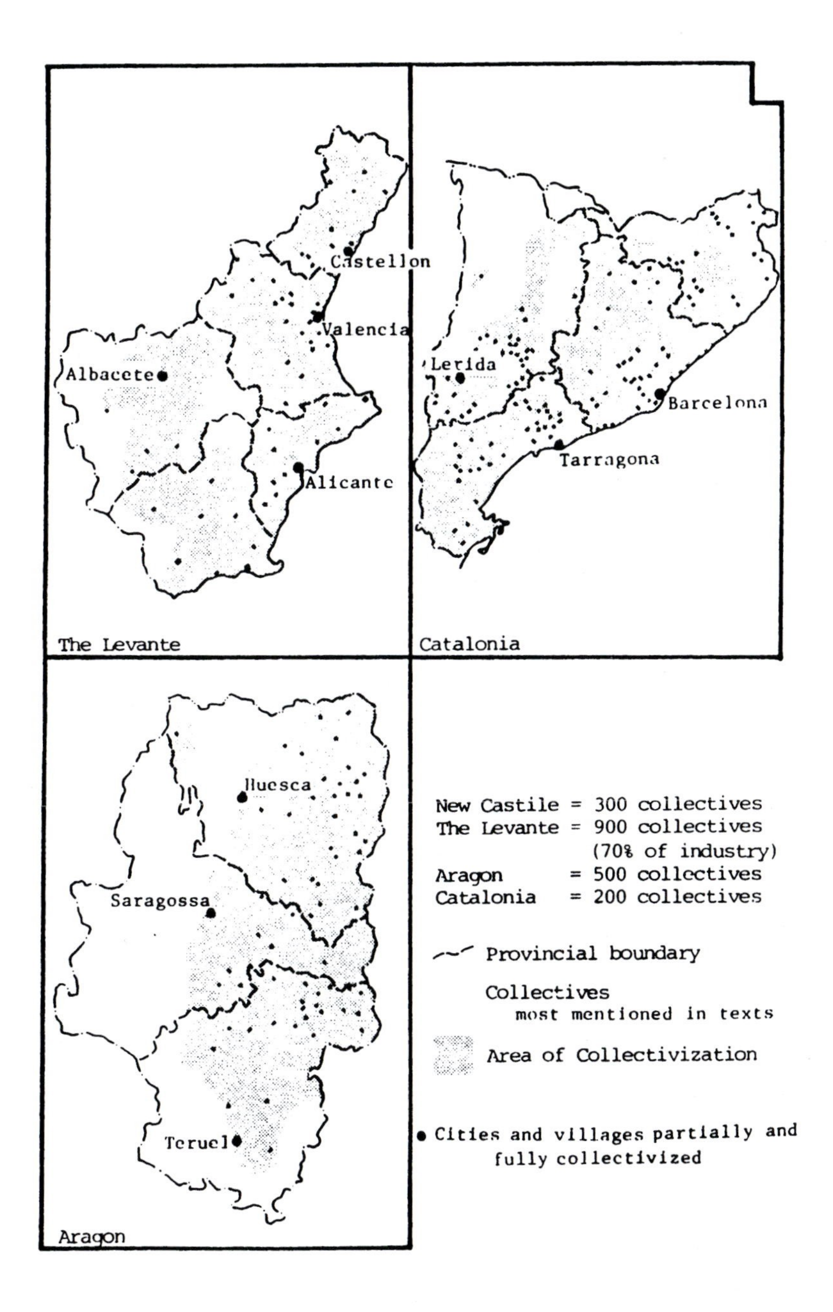
Castellon
Valencia
Albacete
Alicante
The Levante
Lerida
Barcelona
Tarragona
Catalonia
Huesca
Saragossa
Teruel
Aragon
New Castile = 300 collectives
The Levante = 900 collectives
(70% of industry)
Aragon = 500 collectives
Catalonia = 200 collectives
Provincial boundary
Collectives
most mentioned in texts
Area of Collectivization
Cities and villages partially and
fully collectivized

Direct Action: the Development and Spread of Anarchism in Spain Prior to the Civil War

Spain is the only country in the twentieth century where anarchism was adopted extensively as a revolutionary theory and practice. Since the Civil War, many scholars have tried to explain why this particular strain of socialism had such tremendous appeal there. Simplistic explanations which emphasize the isolation or 'backwardness' of Spanish villages, traditions of communalism, and the Spaniard's 'spontaneous talent for cooperation' have proven to be inadequate (Hobsbawm, 1959). Of much greater signficance were a deeply rooted federalist tradition which grew in response to centuries of domination by the State and Church, and the success of working-class organizations in acknowledging the attachment which workers felt to their local areas and the principles of autonomy and self-management. Anarchism in Spain was born out of the tension and contradictions found in nineteenth-century Spanish society; it was also born out of natural human desires for community control.

The Origins of Spanish Anarchism

On the eve of the Civil War, it is estimated that more than one-half of the Spanish population was undernourished, while millions of acres of land were left fallow (Brenan, 1943, xii). To account for this situation, property ownership patterns must be considered.

From the fifteenth century on, nobles in Spain were rewarded with land for loyalty to the Crown. Small farms, irrigation channels, and common arable lands in Castile, Estremadura and Andalusia were replaced by classic *latifundia* (Kaplan, 1976; Costa, 1898). By the eighteenth and nineteenth centuries, a deeply embedded feudal order monopolized economic, social, and political power with 3 percent of the population owning 97 percent of the land (Semo, 1973, 100-28).

The inequities of Spanish life were reflected in the landscape. Most large estate-owners in Spain did not cultivate their land, fearing that irrigation and high yields would motivate the government to redistribute it (Brenan, 1943, 123). Peasants worked seasonally as landless laborers on large estates, living in crowded villages separated by empty countryside. With meagre wages and no access to cultivatable land, large numbers of peasants eventually migrated north in search of food and jobs. On the eve of the Civil War, it is estimated that more than one-half of all factory workers in Barcelona were immigrants from the drought areas of America and Murcia (Brenan, 1943, 124). Northern Spain, in contrast, supported a small semi-capitalist base, producing a few cash crops and small private holdings rented out to tenants as sharecroppers. Most of these rented plots were too small to supply the food needs of a single family, and over time a tradition of cooperative grazing and cultivation developed.

To protect landed interests and the cultural hegemony of Castile, the Madrid government placed a civil guard (*Guardia Civile*) in every *pueblo* and neighborhood, and tried to eliminate all expressions of regional autonomy and peasant communalism. Fearing even the middle class (because of its support for regional autonomy), politicians ignored the pleas of entrepreneurs for monetary support for industrialization (Vicens Vices, 1969). Continued investment in land (symbolic of power), rather than technology, thus 'feudalized' most sparks of capitalist potential in Spain. The Catholic Church, at one time sympathetic to the plight of peasants, also came to share in and support the authority of the feudal state when its lands were threatened with redistribution by a liberal regime in the early nineteenth century. The intense devotion which Spanish peasants later directed toward anarchism, and the intense hatred which they exhibited during the Civil War for the Church, derives at least in part from their earlier religious sense and these feelings of later betrayal (Maura, 1968, 136).

The introduction of a limited capitalist base into the agrarian economy of Spain in the late nineteenth century had prompted the sale of common lands, increasing unemployment among agricultural workers and marking an end to their control over both livelihood and municipal life. The limited introduction of industrial capitalism produced similar effects, wresting control over social and economic life from skilled labor guilds and independent artisans (Kaplan, 1976). These factors, the presence of the *Guardia Civile* in every village and neighborhood, and the social domination of the Church help to explain the anti-authoritarian stance of Spanish workers and their receptivity to the main tenets of libertarian socialism: federalism ('the right of the smallest possible unit to decide its own destiny') and self-management.

Spanish peasants and workers also wanted to believe that revolution was possible *at any time,* and would not have to await the further entrenchment of capitalism. This hope and the attempts of anarchists to supply constructive visions of bottom-up economic and social alternatives go far in explaining the greater appeal of anarchism over Marxism in Spain prior to the Civil War.

The Formal Introduction of Anarchism into Spain

In 1868, soon after Bakunin founded a secret European anarchist organization called the Alliance for Social Democracy, Members Elie Reclus and Guiseppi Fanelli travelled to Spain to explore the 'terra incognitae' of the Spanish working class. The enthusiasm with which their anarchist theories were received by local federalist organizations led to the founding of a Spanish section of the First International (Lorenzo, 1901).

The unique development of anarchist thought in Spain was due in part to social geography. The close identification of people with their local regions led anarchists to form decentralized federations of workers and to pose alternative non-authoritarian uses of space. The three main categories of anarchist thought which emerged in the twentieth century – a European-influenced strain of anarcho-syndicalism, a Spanish version of 'pure' anarcho-Communism, and a compromise between the two – represented nuances in generally accepted principles rather than fundamentally different ideological positions (Elorza, 1972; Breitbart, 1978; Garcia-Ramon, 1979).

European-oriented anarcho-syndicalists focused their writings in Spain on the need for a federation of autonomous working-class organizations to struggle against capitalism (using the tool of the general strike) and to become the basis for a new socialist society (Leval, 1975, 23-5). Refusing to vie for political power, Spanish syndicalists maintained as their ultimate goal the replacement of government with an 'administration' of life through community and workplace networks. The association created in Spain in 1910 to accomplish these tasks was the CNT (Confederation National del Trabajo).

'Pure' anarcho-Communists, inspired largely by the writings of geographers Peter Kropotkin and Elisée Reclus, sought to develop anarchism as a totally new way of life which would affect not just economic behavior but also the relationships between people on a social level (Peirats, 1975, interview; Fredericks, 1972). Anarcho-Communists directed as much concern to cultural, educational, and rural problems as they did to industrial organization. The FAI (Federación Anarquista Ibérica), created in 1927, reflected these broad aims and set out to maintain the ideological purity of the CNT.

Anarcho-syndicalists, who formed a third current of anarchist thought, focused mainly on the issue of alternative work structures and industrial self-management, but adopted their theories to the specific economic realities of Spain in the twentieth century. This group also proposed schemes for the future spatial organization of rural Spain under a communal mode of production (Leval, 1937; Abad de Santillan, 1937).

All three schools concurred on the need to replace structures of authority and feudalistic and capitalist production modes with non-hierarchical federations based upon occupation and home area. It is this emphasis on the importance of social as well as economic values, of 'place' as well as 'function', that most distinguishes Spanish anarchism from other European currents of anarchist thought.

The Diffusion of Anarchist Thought and Practice

When anarchism emerged as a movement in the 1870s in Spain, workers and peasants lacked organization and political consciousness. For most, the road to revolution involved a long arduous struggle to overcome years of powerlessness and indignity.

Anarchist organizations were successful in helping people to transform themselves into a powerful and creative revolutionary force. The principle that revolutionary process need coincide with revolutionary aims led these organizations to construct themselves according to the long-term objectives of federalism, self-management and anti-authoritarianism. Instead of regarding education as a prelude to active involvement, active involvement was seen as a necessary component of political education. Direct action, adopted as a method to initiate radical social change, encouraged people to explore their community and work environments; to examine how centralization dominated the spatial and temporal rhythms of their lives; and to begin to articulate alternatives. This process of self-radicalization influenced the course of collectivization after the Civil War began.

Hundreds of journals, newspapers, and pamphlets spread anarchist ideas throughout Spain in the late nineteenth and early twentieth centuries. They were not intended to lay down a political line. Biographies which captured a sense of the 'whole' person within the diversity of human experience were presented along with ideas for constructive alternatives to capitalism and the State (Leval, 1975, 27-8, 39-40; Gabriel, 1975, interview).

Ateneos – cultural meeting centers which were found in nearly every village and neighborhood in Spain – also played an important role in disseminating radical ideas. These centers contained libraries, lecture and dance halls, and supported a diversity of cultural activities. Because their revolutionary purpose could be disguised, they also became important centers of radical thought when syndicates were outlawed during periods of political repression.

By the end of the nineteenth century, too, itinerant speakers were travelling from village to city, reading newspapers to illiterate workers and peasants, starting night schools, and talking about everything from local problems to social revolution (Peirats, 1977, 177; Hobsbawm, 1959, 190). These *obreros conscientes,* as they were called, did not envision themselves as organizers of a 'proletarian mass.' Rather, by demonstrating a concern for people's immediate needs and by rooting struggle in practical experience, they saw themselves as facilitators of revolution similar to Friere-like educators (Friere, 1970).

In June 1870, the first congress of the Spanish Regional Federation of the First International met to set out its aims: 'We wish the rule of Capital, State and Church to cease and to construct upon their ruins Anarchy, the free federation of free associations of free workers' (Brenan, 1943, 143). Given strong regional feeling, it was decided to work toward these goals by organizing on the basis of occupation *and* locality:

[function & geography]	1. In every locality workers of each trade will be organized in special sections: in addition a general section . . . [for] all workers . . . in trades . . . not yet included in special sections . . . it will be a section of different trades.
[geography]	2. All sections of trade from the same locality will federate and organize a solidary cooperation applied also to matters of mutual aid, education etc., which are of great importance to the workers.
[function]	3. Sections of the same trade belonging to different localities will federate to constitute resistance and solidarity within their occupation.
[geography]	4. Local federations will federate to constitute the Spanish Regional Federations which will be represented by a Federation council elected by the Congresses.

[autonomy]

5. All trade sections, local federations, trade federations as well as the Regional Federation will govern themselves on the basis of their own rules worked out at their congresses.

[participation]

6. All the workers represented by the workers congresses will decide through the intermediary or their delegates, as to the methods of action and development of our organization (Leval, 1975, 20).

Subsequent worker congresses, held in the late nineteenth century by unions affiliated to the International and later by the CNT acting alone, broadened the theoretical base of Spanish anarchism. Local syndicates would hold meetings to establish their positions on a proposed agenda for a congress and then send delegates to espouse these positions. Decisions made at each congress would then be transmitted back and acted upon if local members agreed with their nature and purpose. This communication which persisted between local and regional organizations of the CNT diminished considerably the isolation that once prevented workers from equating their problems.

The major themes addressed at worker congresses included work conditions, the interrelationship between work and community life, alternative forms of work organization based on self-management, and immediate actions to foster revolution. Through an explanation of their current situations and discussions of the above themes, workers were soon able to relate their lack of power and an integrated social and economic life to centrally imposed divisions in time and space – divisions which separated them from each other and from their environments. The importance of self-empowerment and the skeleton of an alternative vision of communal work and community life took shape once these broader relationships were understood.

While congresses provided a forum for the development of anarchist theory in Spain, strikes and *pueblo* seizures challenged central authority directly and provided mini-models for economic and social self-management. To prevent syndicalism from being transformed into ordinary trade unionism, anarchist workers refused to allow the CNT to maintain strike funds or engage in bureaucratic practices which might threaten the autonomy of local work committees. Thus, in spite of an estimated membership of over 1 million by 1936, the CNT never employed more than two paid secretaries and continued to rotate leadership (Leval, n.d., 14; Peirats, 1975, interview).

The aim of an anarchist strike was not to set up mechanisms for collective bargaining or to raise demands which the capitalist system could accommodate. Strikes in Spain often attacked the exploitation of labor by capital directly, seeking worker control over all aspects of the production process. Syndicates also initiated campaigns to gain access to the financial data of firms and encouraged members to learn as much as possible about the production process beyond the narrow confines of their jobs. These measures were used to prove labor exploitation, to help workers overcome the ignorance and alienation which extreme specialization imposed on them, and to prepare for self-management. In the process, traditional capitalist notions of paternalism, hierarchy and efficiency were challenged.

The basic organization of syndicates by locality as well as function encouraged whole communities to seek self-management in their workplaces and living environments through the vehicle of the strike. Supporting a basic anarchist tenet that people struggle not merely as a 'class' but also as human beings who share a common social experience, peasants and workers often used a work strike as an opportunity to seize control over the political organization of their villages and cities. Because struggles for control over work were linked so closely to struggles for control over community, living environments provided another important arena for protest and for the incubation of communal alternatives. These confrontations severely weakened centralized authority, increasing the tensions and contradictions which were generated when actual living and working environments did not fit people's evolving images of an integrated communal way of life. Anarchist organizations built upon these contradictions, encouraging people to avoid a false sense of power and to continually test and expand the limited yet growing power which they possessed.

What Spanish anarchists sought in July 1936 was a massive social revolution which would culminate in the formation of a non-dictatorial form of socialism, recognizing *pueblos,* neighborhoods and workplaces as self-governing units federated through cooperative social and economic networks. The search for an entirely new vision of economy based upon an alternative social vision took place in an atmosphere that expressed considerable tolerance for variety and experimentation. Anarchists were not interested in utopia. The alternative economic and social forms created after 1936 were based on actual models of organization attempted on a limited scale prior to the Civil War.

Just prior to Franco's attack, the Spanish left won a majority in all of the Catalan and Valencian provinces, in parts of Aragon, and most of Andalusia. On 18 July 1936 the power to resist Franco fell into the hands of the most popular leftist organization – the anarchist CNT. Six out of seven of the largest cities in Republican Spain were thus controlled by a revolutionary proletariat. On this day, two revolutions began – one, a well documented military struggle; the other, a social revolution which despite its magnitude and success, history has generally chosen to forget. How would the implementation of an anarchist mode of production be reflected in the Spanish landscape? To what extent would spatial organization play a role in promoting and sustaining social revolution?

The Mode of Production on Rural Anarchist Collectives

The term *collectivization* was employed in Spain after July 1936 to describe many diverse forms of decentralist economic and social organization. These spanned a continuum from limited communalism to true *communismo libertario.* [1] In rural Spain, anarchist peasants established collectives on the land of absentee estate-owners immediately following the attack of the generals in July 1936. Revolutionary administrative committees were also formed by peasants in Catalonia and Levante, where farmers pooled their small land parcels to form, voluntarily, full and partially collectivized villages.

The aim of rural collectivization was to develop a mode of agricultural production which would provide both freedom from hunger and the basis for a classless society.

The Social Organization of Work

Work was of vital importance on a collective. In addition to providing sustenance, it furnished an opportunity for people to display their talents and exercise discretion over their environments. Because work required cooperation, it also provided a nucleus around which new social bonds could be formed. Collectives devoted immense effort to discovering ways to make the whole experience of production a fulfilling one. Bureaucracy, competition, power hierarchies, and all procedures which had a demoralizing effect on workers were avoided. Although delegates were selected on a temporary basis, 'leadership authority' was related to the accomplishment of specific tasks and was discarded as soon as the designated tasks were complete (Souchy, 1937, 151-2).

Anarchist peasants did not believe that efficient agricultural production required hierarchical management and the structures of domination and subordination which characterized feudal and capitalist production modes. The division of tasks on an agricultural collective was therefore designed to fulfill technical requirements without subjecting members to the monotony of minute specialization and without cementing old divisions in status between agricultural and industrial, manual and mental labor.

Elected administrative and work councils with rotating memberships coordinated the economic and social activities on each collective in collaboration with a regional collective council. The assignment of specific jobs on a daily basis was then left to each work team or *brigada,* composed of ten to fifteen people (Leval, 1975, 132; Goldman, 1937a, 3). Each person was encouraged to perform those tasks for which he or she had special competence. Less desirable jobs were rotated or shifted between work teams. These practices and the unification of wages according to need eliminated many of the once-humiliating distinctions that had been made between various categories of work (*Colectividades*, n.d., 37-8, 47-8). No task was accorded any higher status than another and no collectivist was paid for occupying an administrative position (Bouyé, 1964, 114; Souchy, 1937a, 111, 116, 121). Because work was a focal point of communal life and one obvious way to contribute to the well-being of the group as a whole, older members of the collective were also encouraged to volunteer for less strenuous activities (*Colectividades*, n.d., 50). (See Figure 4.2.)

General assemblies, composed of working and non-working members, met monthly to consider economic and social priorities and to evaluate the success of individual productive and social sectors. Agricultural workers, artisans, and factory laborers assigned themselves spokespeople at these meetings to present suggestions about altering work rhythms or redesigning various stages of production (*Colectividades*, n.d., 14; 'Revolutionary Economy', 1937, 3). Numerous factors, including nutrition and demands for food at the front and in cities, were considered when determining what and how much to produce. To facilitate this decision-making process, statistics on agricultural and industrial production were kept by each collective and regional assembly (Souchy, 1937a, 104, 148; *Colectividades,* n.d., 10; Leval, 1975, 95).

Collectives attempting to achieve a greater measure of self-sufficiency relied heavily upon the interdependence of productive sectors. Lucrative sectors were thus encouraged to provide material and physical support to less productive sectors. Intra- and inter-communal work teams

Figure 4.2: Activity Patterns of Peasants Before and After Collectivization

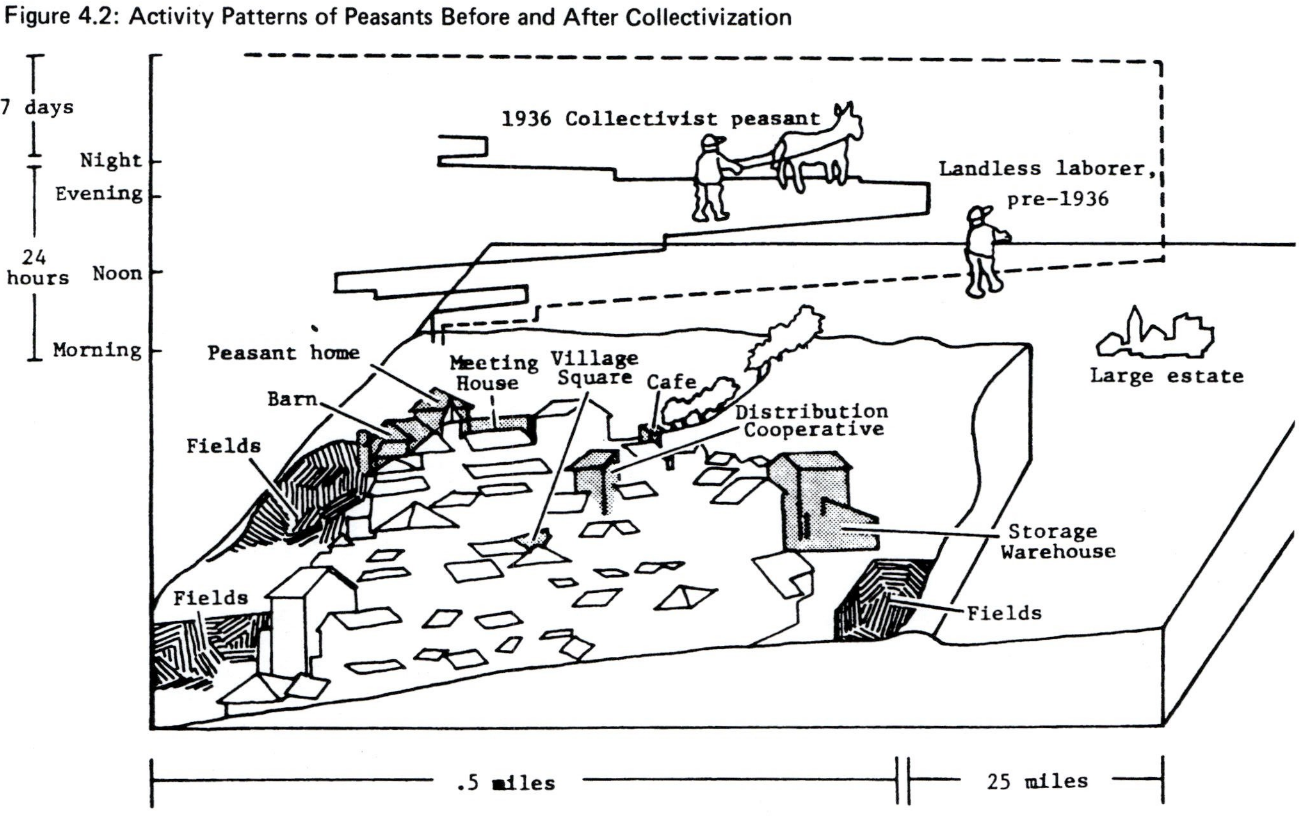

composed of agricultural and industrial workers also exchanged tools, labor, and expertise (Leval, 1975, 318; Souchy, 1937a, 127-8, 136-7; Metcalfe, 1937, 2; *Programme de la Fédération*, 1937, 107). These practices enabled collectives to respond quickly to fluctuating demands for commodities and to obstructions in the flow of supplies caused by the war.

Distribution and Consumption

Collective assemblies provided a democratic arena for planning and for the resolution of conflict. Debates were held continually in an effort to establish accurate measures of economic and social need. Discussions of what and how much to produce were never separated from questions of distribution and consumption. Collectivists were assured that their basic material needs for food, shelter, and clothing would be met and that they could play an active role in determining and fulfilling those needs. This tended to remove much of the psychological basis for acquisitive behavior.

Rural anarchists retained as an ultimate goal the elimination of wages as a measure of the value of labor, and the abolition of money as both value and a determinant of commodity exchange. In practice, a variety of monetary forms developed. Where some form of a wage was retained, it was viewed as a convenient way temporarily to apportion scarce resources and establish exchange with other areas. Most allocations were not made on the basis of hours worked or the quantity of goods produced, but on the basis of family need, as determined by the collective assembly (Souchy, 1937a, 137; Leval, 1975, 152, 190, 304-5; Peirats, 1977, 145; 1943, 113; UGT & CNT Collectives, 1938, 3; Kaminski, 1937, 121-2; Berneri, 1937, 1; De Julio, 1937, 22).

In many parts of Aragon, Levante and Catalonia, collectives replaced official government currency with local coupons or family ration booklets. The buying power of the peseta and the wealth a collective could expropriate from former land- and factory-owners varied greatly by region. Standards of living nevertheless improved in nearly all collectivized areas ('Revolutionary Economy,' 1938, 3; Souchy, 1938, 3; Peirats, 1977, 145-6; 'Collective Farming,' 1937, 3). For those collectives that remained poor, some compensation seemed to come in the achievement of conditions of social equality and in the belief that increased control over one's life would enable future material improvements to be made.

Each rural collective elected special committees composed of consumers and producers to organize the supply and distribution of goods and services from cooperative warehouses. Traditional forms of retailing ceased as villagers came directly to these warehouses for provisions (Souchy, 1937a, 125, 155). Although non-collectivists were allowed to make purchases in cash or kind, they could not buy more in quantity than the quotas set for members. Products that were grown or manufactured locally were generally distributed without charge to collectivists. Shortages of machinery, increased numbers of war refugees, demands at the front, and problems with land that had remained fallow for too long nevertheless required some commodities, like meat, sugar, bread, oil, and wine, to be rationed (Goldman, 1937a; Souchy, 1937a, 164-5; Peirats, 1977, 147).

Technical Knowledge and the Forces of Agricultural Production

In order to increase production, collectives had to expand vastly the use of available resources. After the initial harvest, land-use inventories were made to determine the potential for future production. Loss of European markets and areas cut off by Franco created major problems in regions which had formerly specialized in only one or two crops. Most collectives responded by expanding their storage and preservation facilities, developing new production techniques to improve the quality of crops, and inventing new uses for commodities. For example, collectives in the citrus areas of Levante developed new methods to use the by-products of oranges, fruit pulp, and potatoes (Leval, 1975, 157).

A desire to increase self-sufficiency, decrease the seasonality of labor and maintain access to crops cut off by the war also led rural collectives to diversify production and cultivate formerly vacant land (Leval, 1975, 318-19). Large numbers of animals were added to herds, and linkages were established between various agricultural activities – for example, fruit-growing, bee-keeping, and honey production (Souchy, 1937a, 145; Leval, 1975, 146, 169). Reflecting an attachment to place, many collectives exhibited a concern for long-term conservation, rotating crops, planting trees to prevent soil erosion, and establishing laboratories to research new planting techniques and animal waste fertilization (*Colectividades*, n.d., 2; Bouvé, 1964, 108-19; *Plano Regional*, 1937, 36).

In an attempt to increase the self-sufficiency of regions, anarchists introduced new industries into rural areas (Souchy, 1937a, 122-3). Inventories of industrial needs and potential were taken to determine which manufacturing activities to introduce anew, which to expand, and which to dissolve. Because many industries faced raw material shortages, new production processes had to be developed that employed materials which were easier to obtain. Levante collectives concentrated on developing industries for fruit and vegetable processing and for the distilling of alcohol, juices and perfumes (*Congresso Regional de Campesinos de Levante*, September 1936, 222). Industries which utilized the waste products of other industries were also expanded (*Colectividades*, n.d., 8; Souchy, 1937a, 148).

In spite of the disruptions caused by war, most rural collectives were able to modernize and expand agricultural production. New and existing technologies were adopted for the first time in centuries to meet Republican Spain's food and material needs (Leval, 1975, 182, 188).

Transformations in Rural Spatial Organization

Spatial organization had played an important role in inhibiting mutual aid, equality, and social justice in Spain prior to 1936. During the Civil War, rural anarchists set out to transform their natural and built environments to accommodate communal economic and social priorities.

Village landscapes acquired a considerably new character due to the social relations and technical organization of communal production. Membership in larger federations of collectives encouraged modifications in transportation networks, circulation patterns, and administrative boundaries. A natural and built environment which once reflected the tensions and contradictions of feudal and capitalist modes of production began to reflect a certain harmony.

The Communal Economic Landscape

Spatial organization performed a key role in disseminating anarchist ideas in Spain prior to the Civil War. This was apparent in the federal structure of syndicates and in the efforts of peasants to assert village autonomy. After collectivization, assemblies met immediately to consider the question of land use.

Since top priorities after the harvest were to diversify planting and to bring as much unproductive land under cultivation as possible, irrigation was required. Many collectives set out to build reservoirs, water conduits, bridges and wells in areas which had lacked water since the thirteenth century (Berneri, 1937; Tambaret, 1938, 2; Souchy, 1937a, 145; *Colectividades,* n.d., 49-50; Parker, 1938). The acquisition of water resources enabled them to uproot crops like olives, and prepare the land for other more necessary forms of cultivation. Estimates suggest that collectivized villages harvested up to five times as many crops as before the Civil War (Peirats, 1977, 140; Souchy, 1937, 113, 143; Tambaret, 1938, 2).

The introduction of many small industries on collectives also required the construction of new barns, storage facilities, mills, and processing centers (Souchy, 1937a, 122; 'Revolutionary Economy,' 1937, 3). Small scattered workshops were often consolidated and their functions integrated within a single new building or a converted church (Peirats, 1975, interview; Souchy, 1937a, 122-3, 140). Because of their interior spaciousness and centrality, churches were also used as distribution warehouses, after tiled floors, water pipes, partitions, and new windows were installed (Marcos-Alvarez, 1972, 131; Broué and Témime, 1970, 150).

The Communal Social Landscape

A contingent of peasant guards adorned in revolutionary attire was a common sight as one approached a collectivized village: 'picturesque, [They] look[ed] as if they were cut out from a Goya painting: peasant clothes . . . but . . . with red or red-black neckties . . . distinguished from ordinary mortals by red badges with the stamp of their [political] organization' (Borkenau, 1937, 93). Moving from the outskirts of the village toward the center one was also likely to notice posters or signs adorning the façades of shops that read 'Tailors' Collective No. 1,' 'In this firm one works collectively,' or 'Villa Kropotkin' (Souchy, 1937a, 160). Rural collectives were not, of course, constructed anew. Formed within older villages, the existing built environment had to be altered to accommodate new priorities. Even structures which retained the same outward appearance after collectivization were markedly transformed in function (Figure 4.3).

Most collectives placed emphasis on cultural expression, scheduling musical events and movies at no charge, and placing on display works of art from the church or homes of the rich (Borkenau, 1937, 134; 'Revolutionary Economy,' 1937, 3; Souchy, 1937a, 125-6; Berneri, 1937, 1). The desire for a vibrant cultural life reflected itself in the built environment. *Ateneos* (centres of study and cultural activity) which had attained political importance on the social landscape before July 19, were expanded to include more books, and lecture and dance halls (Souchy, 1937a, 144, 162; Gabriel 1975, interview; 'Revolutionary Culture,' 1938, 3). Collectives also made an effort to integrate cultural activities into daily work routines, allowing dancing and café visits to break up the tedium of the working day (Thomas, 1971, 252-3). One collectivized textile mill near Valencia went so far as to install a dining-room, lecture hall, library, and radios on shop floors (Goldman, 1937c).

Before the Civil War, few schools were provided for towns with less

Figure 4.3: Changes in the Built Environment After Collectivization

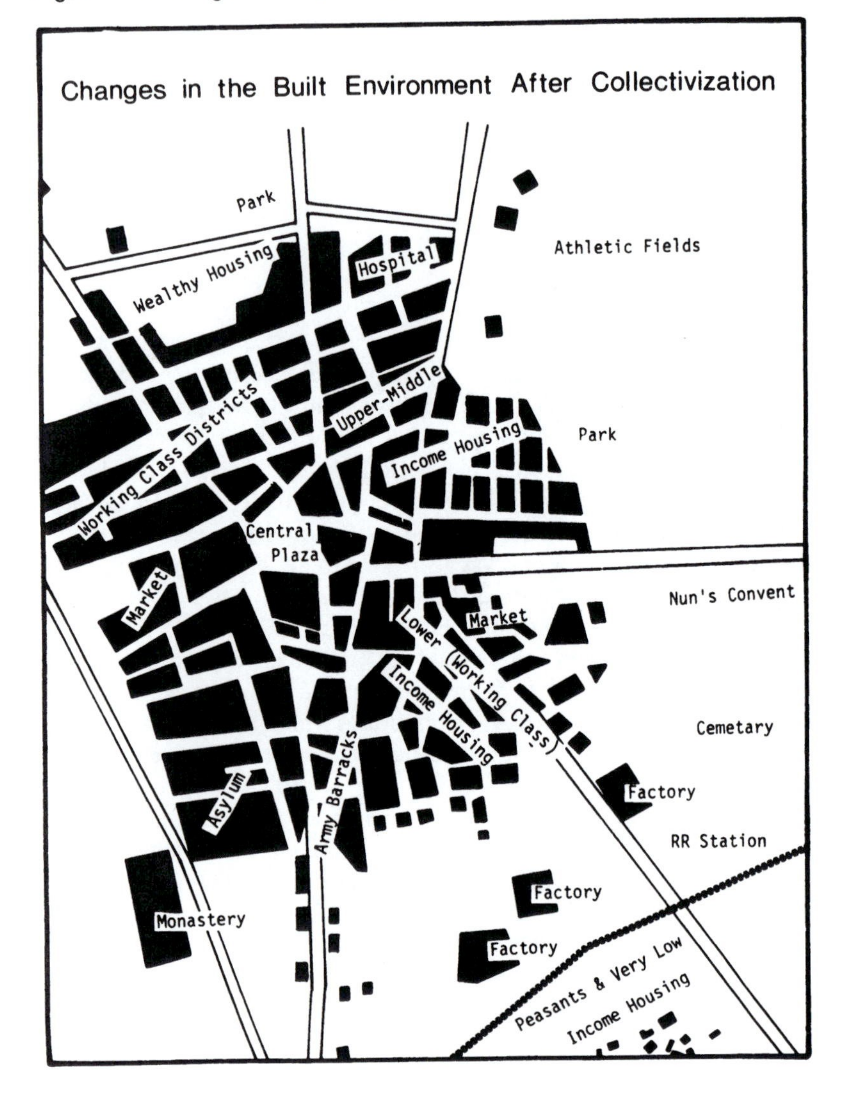

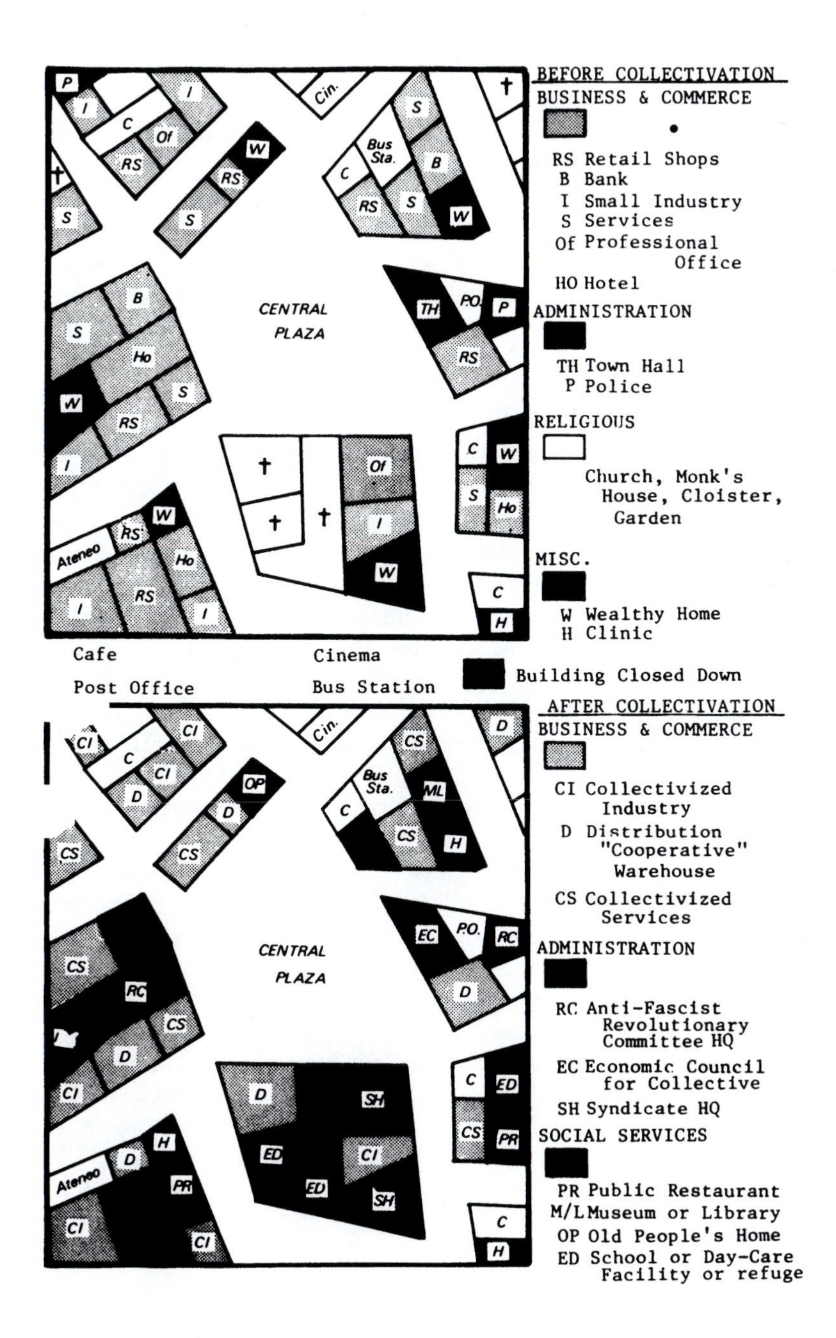
BEFORE COLLECTIVATION
BUSINESS & COMMERCE
RS Retail Shops
B Bank
I Small Industry
S Services
Of Professional Office
HO Hotel
ADMINISTRATION
TH Town Hall
P Police
RELIGIOUS
Church, Monk's House, Cloister, Garden
MISC.
W Wealthy Home
H Clinic
CENTRAL PLAZA
Ateneo
Bus Sta.
Cin.
P.O.
Cafe
Cinema
Post Office
Bus Station
Building Closed Down
AFTER COLLECTIVATION
BUSINESS & COMMERCE
CI Collectivized Industry
D Distribution "Cooperative" Warehouse
CS Collectivized Services
ADMINISTRATION
RC Anti-Fascist Revolutionary Committee HQ
EC Economic Council for Collective
SH Syndicate HQ
SOCIAL SERVICES
PR Public Restaurant
M/L Museum or Library
OP Old People's Home
ED School or Day-Care Facility or refuge
CENTRAL PLAZA
Ateneo
Bus Sta.
Cin.
P.O.

than 5,000 people, and approximately 70 percent of the rural population was illiterate (Leval, 1973, 158). During the Civil War, more money was allocated to education in these communities than to the military, government, or agriculture. The only higher expenditure was public works ('Education and War,' 1937, 1).

Anarchists had always considered education a powerful tool for radical social change. Schools and libraries, rarely seen in the Spanish countryside before collectivization, thus became prominent features on the social landscape after July 1939. In Levante, most collectives in the regional federation (numbering 900) have their own school by 1938 (Dolgoff, 1974, 168; Souchy, 1937b, 47; Berneri, 1937, 1; *Colectividades,* n.d., 10, 43).[2] Efforts were also made to integrate learning into daily work routines by making courses available to workers and by encouraging children to venture out of the classroom to explore the various activities taking place on the collective (*Colectividades*, n.d., 43; 'Anarchism by Education,' 1937, 3).

Rural Spain suffered acute health problems prior to the Civil War due largely to the inequitable spatial distribution of health services. During the war, high casualty rates multiplied the pressures on already over-burdened facilities. Collectives attempted to alleviate these pressures by sending mobile surgical teams to the front. Collective doctors organized anti-typhoid and vaccination campaigns, purified water, and oversaw the construction of new hospitals and pharmaceutical laboratories (Larrut, 1938, 4; Metcalfe, 1937, 2). Several elderly people's homes and children's colonies were also established in areas removed from the fighting.

The poverty of many rural collectives prevented them from investing all they would have needed to bring their health and service facilities up to standard. The provision of social services thus became an important function of the district, provincial and regional federations of collectives. The National Federation of Public Health, a section of the CNT, had 40,000 members by 1937, including some of Spain's most active militants. In February 1937, local CNT syndicates sent representatives to a National Congress in Valencia where the plans were laid to link local health activities and reorganise the training of public health workers (Leval, 1975, 275). The greatest emphasis was placed on setting up dispensaries to track down and monitor disease in rural areas.

Regional federations of collectives also helped to bring doctors together and collect information on disease. One example was the Sanitary Services syndicate which divided Catalonia into nine large health sectors, and several secondary sectors for the purpose of allocating services.

Although a committee in Barcelona helped to coordinate the network, district councils maintained direct touch with individual collectives and played an important role in responding to their unique medical needs (Montseny, 1975, interview).

The Regional Economic Landscape and Collective Exchange

Rural collectivization did not create small, inward-looking, totally self-sufficient entities, nor did it eliminate the need to seek solidarity in exchange with other areas. The limited degree of self-sufficiency in necessities which most rural collectives achieved through intra-communal linkages between agriculture and industry provided some defense against external domination and dependency. Collectives then found they could exchange with each other and participate in cooperative planning on a more equal basis. Alterations in land use which were made within individual villages eventually extended out into the regional landscape as a result of these external linkages.

The informal exchange of products and expertise began between rural collectives as an expression of mutual aid during the months following the July uprising. Food and supplies were sent to poorer collectives and to local industries and their workers ('Revolutionary Economy,' 1936, 3). Industries, in turn, provided information on planting and crop processing (*Plano Regional*, 1937, 29-30). In many parts of Aragon, collectives also shared supply depots. Instead of maintaining strict records on exchange, participants drew what they needed from these depots and then contributed to them in amounts roughly equivalent (Souchy, 1937a 161).

An additional goal of rural collectivization was to overcome traditional animosities between agricultural and urban-industrial workers (*Tierra y Libertad*, 20 March 1937). This was accomplished in part by replacing competitive exchange between 'unequals' with mutual association for the benefit of both rural and urban areas. Peasants linked the problems of cities not to a distinct 'urban personality' but to the particular contradictions and inequities inherent in industrial capitalist production. Recognizing the relationship between economic centralization and urban expansion, they also discussed the role that cities played in extracting surplus from hinterlands (Alaiz, 1937, 7). Were cities to relinquish this economic and social 'dictatorship', it was felt that urban and rural linkages could provide a solid basis for collective strength ('Social Aspirations,' 1936, 3).

The social revolution began to bridge the gulf between peasants and city workers, and to reduce the historical flow of surplus value toward

Figure 4.4: Generalized Scheme for Regional Collective Exchange

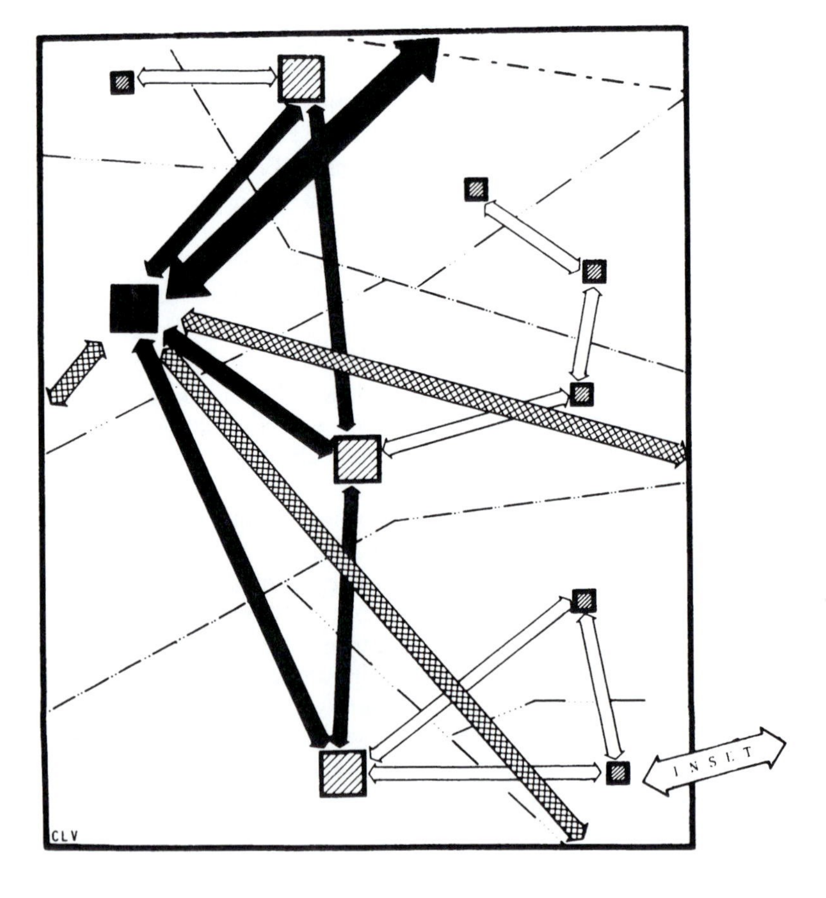

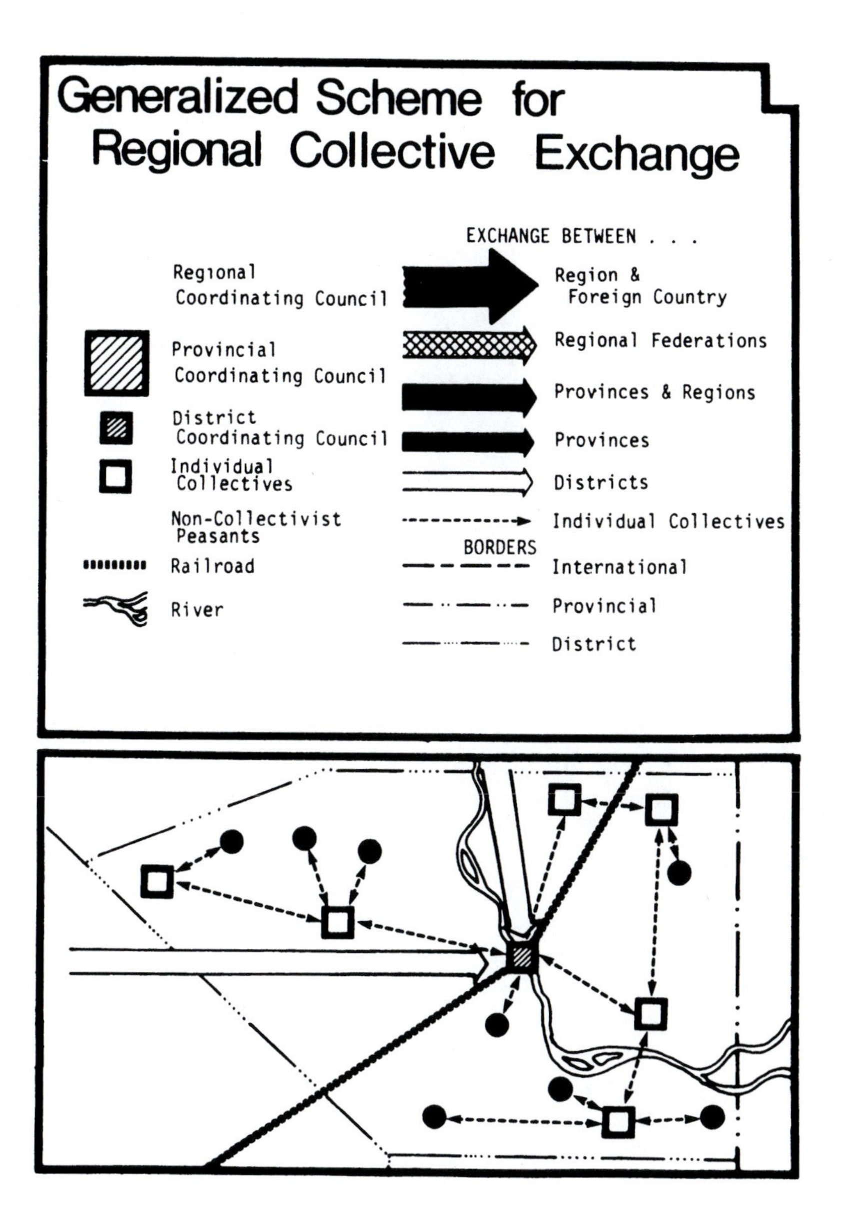
Generalized Scheme for Regional Collective Exchange
EXCHANGE BETWEEN . . .
Regional Coordinating Council
Region & Foreign Country
Provincial Coordinating Council
Regional Federations
Provinces & Regions
District Coordinating Council
Provinces
Individual Collectives
Districts
Non-Collectivist Peasants
Individual Collectives
BORDERS
Railroad
International
River
Provincial
District

cities. The most important linkages were established in the direct transfer of food by 'bread committees' in rural communes to urban neighborhoods (Borkenau, 1937, 183-4; Souchy, 1937a, 164). Urban syndicates provided rural collectives with technical help on water resource projects, labor for the harvests, and medical aid. Collectivized industries also received fruit and vegetables in exchange for machinery and manufactured products (Souchy, 1937a, 46; Berneri, 1937; *Plano regional,* 1937). Only when the Republican government began to requisition or confiscate produce from rural collectives without reimbursement did antagonisms re-emerge between the city and the countryside (Oehler and Blackwell, 1937). This happened with greater frequency as the Communist counter-revolution against the anarchists and collectivization progressed (Marcos-Alvarez, 1972).

Regional Federations of Collectives

Spanish anarchists posed the idea of a 'region' as the most fundamental cell in economic and social life. Composed of villages, districts, and provinces, each region was to embody cultural and ecological traditions. During the Civil War, the combined economic power of rural collectives in regional federations also protected the autonomy of individual villages by aiding them to overcome severe wartime pressures.

Several large regional federations of collectives were formed between July 1936 and June 1937. The most successful of these were the Regional Federation of Peasants of Levante, the Regional Federation of Peasants of Castile and the Aragon Council. Although their regional organizations each displayed a different strength (production, consumption, or distribution), these federations helped to facilitate the exchange of commodities within and between collectivized districts. Figure 4.4 provides a generalized scheme for this exchange process.

Beginning at the local level, administrative *delegados* in each collective submitted reports on imports and exports to a record-keeper for the region. Surplus items (as defined by the collective) were then turned over to the *caja de compensacion* of the district. This coordinator was usually located in the largest village, the one with the strongest libertarian roots, or the one most accessible to others. The task of the district council (composed of representatives from each collective) was to help redistribute surplus within the district or use it to acquire additional commodities from within and outside the larger region (see insert of Figure 4.4).

Over 900 collectives were formed in Levante after July 1936. These lasted until the end of the war due to their distance from the front.

This number encompassed more than 40 percent of the population and more than three-quarters of the land area in the region. Collectives produced 50 to 60 percent of Levante's products and served 70 percent of its distribution needs (Broué and Témime, 1970, 159). The Regional Federation of Peasants of Levante became most noted for its intricate patterns of distribution, and for its concern with scientific innovation. The most important distribution networks were those set up to supply the regional and international market with oranges.[3]

The success of Levante collectives in coordinating their activities of production and distribution depended largely on the rigorous accounting procedures which they employed. Each collective kept detailed statistics on productive potential and consumptive need. A special section for Advice and Statistics in Valencia met regularly with representatives to pass on the information from other parts of the region (*Federacion Regional de Campesinos de Levante*, 1937).

Research teams were also organized by the Federation to advise collectives on housing rehabilitation and the most productive uses of agricultural space given travel time to, and climate within, purchasing areas. The proximity of most collectives enabled intercommunal work teams to collaborate on curbing plant disease, pruning vegetation, and transporting products to market or the front. Thousands of peasants from Levante also journeyed to Castile and Catalonia to provide collectives there with information on new agricultural techniques (Marcos-Alvarez, 1972, 123).

Collectivization also proceeded rapidly in Republican-controlled areas of Castile. By 1937, approximately 300 collectivized villages and 700 syndicates were affiliated with the Regional Federation there (Leval, 1975, 182-3). The main task of the Federation was to integrate the agricultural and industrial economies of Castile and to synchronize production and distribution in collectivized zones. This was a formidable task, as most of the area was under constant Fascist assault.

Perhaps the most outstanding creations of the Regional Federation of Peasants of Castile were District Equalization Funds which served to redistribute in the form of products what little wealth there was. Statistics indicate that a considerable spirit of cooperation prevailed, as the regional headquarters received and distributed more than 11 million pesetas in a single year (Leval, 1975, 185).[4] The Federation in Castile also helped poor collectives solve agricultural problems by expanding research on improving crop yields and by encouraging small industries to locate in rural areas (Parker, 1938, 3). The spread of technical information was facilitated by the publication of professional anarchist agricultural journals such as *Campo Libre.*

Collectivization in Aragon proceeded swiftly and in a more radical fashion than anywhere else in Spain. The Council of Aragon was formed in the early fall of 1936 to facilitate production and exchange among collectives and to protect them from exploitation by Catalan businessmen and non-anarchist militias. At its height, the Aragon Federation included 450 collectives and more than one-third of the population living within Republican parts of the region (Lorenzo, 1969, 152).

The first task of the Aragon Council was to regroup collectives into a coherent ensemble by altering slightly the traditional boundaries between villages. Collectives then formed their own contiguous District Federations. Efforts were made to safeguard the rights of non-collectivists in the process (*Programme de la Fédèration*, 1937, 108). Redistribution schemes similar to those described earlier were also established, as was a common fund to facilitate exchange between Aragon and other regions and foreign nations. Towns which were located near regional borders played an especially important role in this process.[5]

Aragon collectives contributed enormous amounts of food to the war front and often experienced internal shortages as a result. Government harassment of collective exchange networks was, however, more acute in Aragon than anywhere else. A brutal anti-collectivist campaign, initiated in October 1936 by Vincent Uribe (Communist Minister of Agriculture for the Republicans), culminated in a military assault on Aragon collectives in August 1937. The atrocities that occurred as a result of this campaign are well documented in political texts on the Civil War (Chomsky, 1969; Lorenzo, 1960; Brademas, 1955; Prats, 1938; Bolloten, 1961). Peasants were shot in the fields, badly needed food was destroyed, rural hospitals were burned, and all efforts at urban-rural exchange were suppressed (Marcos-Alvarez, 1972). By late 1937, the Aragon Council and its members were completely demoralized (Peirats, 1977, 259). Last-minute efforts on the part of the government to revive agricultural production and prevent mass starvation in cities failed.

The Collective Regional Space-Economy

Networks of exchange established by Regional Federations before 1938 promoted significantly new patterns of circulation in Spain, penetrating the artificial barriers which once isolated one rural village from another. Boundaries between areas were altered, public works projects begun, and transportation systems revamped.

Prior to 1936, boundaries between villages, districts and provinces were drawn by the Madrid government to discourage regional autonomy. In August 1936, the autonomous government of Catalonia implemented a new set of territorial demarcations based on an earlier perception study by geographers which attempted to get a sense of the 'real' areas with which people identified (*Generalitat de Catalunia*, 1933; Gasch, n.d.). Similar territorial alterations were made in other anarchist regions to facilitate inter-communal exchange and to acknowledge the emotional attachment which peasants had to their local environments.

Backward and forward economic linkages established between collectives and the need to move troops and supplies created a demand for improved accessibility, including new roadways, bridges, and telegraph wires (Metcalfe, 1937, 2; Louzan, 1937, 3). Massive alteration of major transport lines in Republican Spain began on 24 July 1936 when the National Syndicate of Railways (mainly anarchist) initiated worker takeovers of the main rail lines in Catalonia. Workers also seized the main lines on the Madrid-Saragossa-Alicante (MSA) route. Top managers were dismissed and replaced by worker-technicians. Special revolutionary committees selected by CNT and UGT (Socialist) workers guarded stations from attack and coordinated the 292 trains operating per day between Castile, Aragon, and Levante (Souchy, 1937a, 34-6). Eventually, a single Federation which brought together the Catalan, Northern and MSA networks was formed.

Anarchist transport workers collected data on existing services and potential areas of expansion via questionnaires distributed at stations (Leval, 1975, 262-3). On the basis of this information, several schemes were proposed to eliminate the unnecessary duplication of labor and resources encouraged by former competition between private lines. Redundant rail and sea links between well serviced areas were halted. Materials and labor from these lines were then used to construct or improve passenger facilities in isolated rural districts. They were also used to service the war front (Souchy, 1937a, 37).

Conclusion

Franco's attack on the Republican government in July 1936 provided Spanish peasants with the first real opportunity to seize land and implement communal anarchist principles on a large scale. A sequence of change was set in motion which originated with social relationships, proceeded to economic and technical organization, and ended up promoting large-scale alterations in the use of both time and space. A social revolution which began by initiating change *in people* thus ended up creating a *significantly new environment.*

Prior to 1936, Spanish peasants lived a contradiction. Daily routines were split temporally, spatially and experientially between work – over which there was little control – and social life, over which there was limited control and some personal reward. Capitalist and feudalistic interests dominated the time and space rhythms of work directly. These interests also indirectly affected the social routines of village life, making it difficult for peasants and workers to satisfy their common health, educational and cultural needs.

Communal proprietorship of the land and the elimination of class in anarchist areas after July 1936 replaced private land ownership and capitalist or feudal hierarchies. An administrative apparatus which called for the participation of all working members of a collective supported these changes and altered productive relations. Workers shared responsibility for decision-making and coordination of jobs. Fundamental changes in the social relations of production also encouraged cooperation between various branches of agriculture and industry, and between the city and the countryside.

Control over the means of agricultural production led to the expanded use of resources and altered farming techniques. The combined changes in social and technical relations were then supported by sizable alterations in the use of space at the local and regional levels. Peasants diversified planting and built irrigation systems which brought land under cultivation that had gone untilled for generations. For the first time in centuries, the problem of starvation was reduced. Concern for local area and autonomy did not close horizons in anarchist Spain – it opened them. While individual collectives aimed for self-sufficiency in basic necessities to decrease outside dependence, they also established strong inter-communal ties.

A communal mode of production also affected the content and meaning of social life, enabling peasants to absorb libertarian principles directly into daily social routines. The tensions created by war and an incredibly demanding work schedule would seem to have precluded peasants from deriving a sense of personal satisfaction from this communal existence. The fact that they perceived their lives as *less* tedious, however, indicates that tedium was as much a product of mental stress and personal contradiction as of physical duress. Prior to collectivization, much stress came from having to live two contradictory lives – a secretive communal social life, and a competitive economic existence marked by the inability to gain control over the means of production. Collectivization enabled workers to escape oppressive authority and exercise personal prerogative over their economic and social lives. This humanization of economic and social values spawned new rhythms of life and generated the energy necessary to lessen the impact of the overwhelming problems that remained.

The seizure of control over life by anarchist peasants in 1936 thus buried much of the exhaustion that went with an old way of life. For a short period of time, at least a part of Spain was able to develop a mode of production which supported rather than contradicted the behavior that people valued in their social lives. Anarchism was removed from Spain in 1939 after Franco, aided by Hitler and Mussolini, achieved military victory. Thousands of militants went each day before the firing squads in what has been described as 'a continual orgy of murder, a St. Bartholomew's Eve that lasted for six years' (Peirats, 1977, 342).

If we adopt Malatesta's view (1891) and refrain from measuring the success of revolutions solely in terms of their ultimate political victory, it is clear that Spanish anarchists have a number of constructive achievements to which they can point. Significant, lasting improvements were made in the rural economic landscape as a result of collectivization. Although not discussed in this chapter, the scale and success of urban collectivization and workers' control of industry were also great, impressing 'even highly unsympathetic observers' (Chomsky, 1969, 78). Anarchists demonstrated the importance of viewing both the workplace and the community as terrains for struggle and as places where cooperative relationships could be revived. Because radical social change was not initiated by hierarchical elitist organizations, issues emerged from daily life, and people were more readily able to transform themselves into active agents for change. Industrial syndicates and agricultural collectives redefined the concept of efficiency, achieving more productive forms of organization than firms in either the private or nationalized sectors.

Anarchist decentralism was more, however, than a social process and spatial form: it was based upon a significantly new philosophy of life. Its characteristics – self-management, integration of economic and social activities, smallness of scale, and federalism – were aimed at developing positive human qualities. That Spanish anarchism succeeded in doing this is indicated by the personal traits of its adherents – many people who were anxious to cooperate with others, who understood the needs of others and responded, who maintained interest and vitality in their work in spite of long hours, who lived close to and respected nature, and who managed their lives responsibly without entrenched forms of authority.

Although anarchists suffered political and military defeat in Spain in 1939, it was not without contributing something concrete for contemporary social activists to build upon – the practical achievements of collectivization, the integration of a community and environment, and an understanding that the root of personal freedom lies in social action.

Notes

1. *Communismo libertario* was defined by the CNT in February 1937 as

> the organization of society without a state and without private property . . . The centers of organization . . . are the Syndicate and the free municipality . . . These two organizations, federative and democratic . . . have sovereignty over their own decisions . . . [they] will take collective possession of everything that now belongs to the sphere of private property. They will regulate production and consumption in every locality, although they will leave people in charge of their own actions . . . liberatarian communism . . . makes compatible the satisfaction of economic necessities and respect for . . . Liberty . . . egoism is . . . replaced by the broadest social love *(Boletín de Informacion,* 193, 27 February 1937, 4).

2. If funds were not available to construct new buildings, convents, churches, or old army barracks were converted to serve as classrooms. Anarchists also tore down many walls and enlarged windows in existing schools to provide an atmosphere they felt was more conducive to learning and free expression (Souchy, 1937a, 135, 146; Goldman, n.d., 4).

3. More than half of the total orange production in the region was handled by the peasants' Federation during the war. Of this total, 70 percent were sold in Europe (Broué and Témime, 1970, 158).

4. As an indication of how much this meant in produce, 225 pounds of corn sold for only 58 pesetas (Leval, 1975, 186).

5. Binefar, a town of 5,000 within a district of 28 collectivized villages, handled more than one million pesetas' worth of commodities in exchange with Catalonia and other Aragon collectives between October and December of 1936 (Leval, 1975, 117).

References

Abad de Santillan, Diego, 1937. *After the Revolution· Economic Reconstruction in Spain Today.* New York: Greenberg.

Alaiz, Felipe. 1937. *Por una economia solidaria entre el campo y la ciudad.* Barcelona; Oficinas de Propagande CNT-FAI.

'Anarchism by Education: Self-Development Only Through Freedom.' 1937. *Spain and the World, 1* (8) (19 March), 3.

Berneri, C. 1937. 'Revolutionary Economy: Two Collectivized Villages in Catalonia,' *Spain and the World, 1* (17) (11 August), 1.

Boletín de Información CNT-AIT-FAI. 1936-8. Barcelona.
Bolloten, Burnett. 1961. *The Grand Camouflage.* New York: Praeger.
Borkenau, Franz. 1937. *The Spanish Cockpit.* London: Faber and Faber.
Bouyé, Jacques. 1964. 'L'homme dans l'industrie.' In *Problèmes Contemporains*, edited by Groupe Humaniste Libertaire. Paris: Librairie Publica.
Brademas, Stephen. 1955. 'A Note on the Anarcho-Syndicalists and the Spanish Civil War,' *Occidente, 11* (2), 121-35.
Breitbart, Myrna, M. 1978. 'The Theory and Practice of Anarchist Decentralism in Spain, 1936-39: The Integration of Community and Environment,' PhD dissertation, Clark University, Worcester, Mass.
Brenan, Gerald. 1943. *The Spanish Labyrinth.* London: Cambridge University Press.
Broué Pierre, and Témime, Emile. 1970. *The Revolution and the Civil War in Spain.* London: Faber and Faber.
Carr, Raymond (ed.). 1971. *The Republic and Civil War in Spain.* London: St Martin's Press.
Chomsky, Noam. 1969. *American Power and the New Mandarins.* New York: Vintage.
Colectividades de Castilla. n.d. *El colectivismo en la Provincia de Madrid,* ediciόners de la Federatión Regional de Campesinos y Alimentación del centro. CNT-AFT.
'Collective Farming.' 1937. *Spain and the World, 1* (2) (19 May), 3.
Costa, Joaquin. 1898 (1944). *Colectivismo Agrario en Espana.* Buenos Aires: Editorial Americales.
Cultura y Acción. 1936-8. Alcaniz. Organo de la Regional de Aragón, Rioya y Navarra, FAI.
De Julio a Julio: Un Ano de Lucha. 1937. Barcelona: Tierra y Libertad.
Dolgoff, Sam. 1974. *The Anarchist Collectives: Worker Self-Management in the Spanish Revolution.* New York: Free Life.
'Education and War.' 1937. *Spain and the World, 1* (7) (5 March), 1.
Elorza, A. 1972. 'Une Conception Scientifique du Communisme Libertaire: D.A. de Santillan,' *Autogestion et Socialisme, 18-19* (January-April), 83-102.
Federación Regional de Campesinos de Levante. 1937. *La Administratión en el Campo,* CNT-AIT Valencia.
Fredericks, Shirley Fay. 1972. 'Social and Political Thought of Federica Montseny, Spanish Anarchist, 1923-7,' PhD dissertation, University of New Mexico.
Friere, Paulo. 1970. *Pedagogy of the Oppressed.* New York: Herder and Herder.
Gabriel, P. Interview, 8 July 1975, Barcelona, Spain.
Garcia-Ramon, Dolores. 1979. 'The Shaping of the Rural Landscape: Contributions of Spanish Anarchist Theory,' *Antipode, 10-11,* 71-82.
Gasch, Emili and Roca, Francese. 1973. 'La Nova Economia Urbana 1936-9,' *Cuadernos de architecture y urbanism, 22* (Nov.-Dec.), 66-78.
Generalitat de Catalunya. 1933. *Divisio Territorial: Estudia: Projectes Nomenclator de Municipis.* Barcelona.
Goldman, Emma. n.d. 'Lure of the Spanish People.' Ann Arbor, Michigan: University Microfilm Service, 63-1624.
—, 1937a. 'Albalate de Cinca: A Collectivized Village,' *Spain and the World, 1* (7) (5 March), 3.
—, 1937b. 'My Second Visit to Spain, Sept. 16 – Nov. 6, 1937,' Archives: Federicao Arcos.
Goodman, Paul, 1965. *People or Personnel.* New York: Random House.
Hobsbawm, E.J. 1973. *Revolutionaries: Contemporary Essays.* New York: Pantheon.
Kaminski, H.E. 1937. *Ceux de Barcelone.* Paris: Les Editions Denoil.

Kaplan, Temma. 1976. *Anarchists of Andalusia.* New Jersey: Princeton University.
Larrut. 1938. 'A Doctor in Aragon,' *Spain and the World, 2* (29) (2 Feb.), 1, 4.
Leval, Gaston, n.d. 'Kropotkine et Malatesta.' Paris: Sofrim, s.a. 8pp.
—, 1937. *Precisiones sobre el anarquismo.* Barcelona: Edition Tierra y Libertad.
—, 1971. *Espagne Libertaire 1936-9.* Paris: Editions du Cercle.
—, 1975. *Collectives in the Spanish Revolution*, translated by Vernon Richards. London: Freedom Press.
Lorenzo, Anselmo. 1901 (1947). *El Proletariade Militante.* Toulouse, France: Editorial del Movimento Libertario Espanol – CNT.
Lorenzo, Cesar M. 1969. *Les anarchistes espagnols et le pouvoir, 1869-1969.* Paris: Editions du Seuil.
Louzon, R. 1937. 'puigcerda,' *Spain and the World, 1* (7) (11 August), 3.
Malatesta, Errico. 1891. *Anarchy.* London: Freedom Press.
Marcos-Alvarez, Violett. 1972. 'Les Collectivités espagnoles pendant la révolution (1936-1939),' *Autogestion et Socialisme, 18-19* (Jan.-Avril), 119-42.
Maura, J. Romero. 1968. 'Terrorism in Barcelona and its Impact on Spanish Politics 1904-1909,' *Past and Present, 41* (Dec.), 130-83.
Metcalfe, H. 1937. 'A Canadian Socialist's Views on Conditions in Catalonia,' *Spain and the World, 1* (12) (19 May), 2.
Montseny, Federica. Interview, 18 July 1975, Toulouse, France.
Oehler, Hugo and Blackwell, Russell. 1 May 1937. Letter written to NYC Socialists from Barcelona. Spanish Collection, Brandeis University Library.
Orwell, George. 1938. *Homage to Catalonia.* London: Penguin.
Parker, Edgar. 1938. 'Revolutionary Economy: The Agricultural Collectives Complete Their Work,' *Spain and the World, 2* (39) (26 August), 3.
Peirats, José. Interview, 10 July 1975, Montady, France.
—, 1977. *Anarchists in the Spanish Revolution,* translated by Mary Slocombe and Paul Hollow. Detroit: Black and Red.
'Pleno Regional de Sindicatos de Aragon, September 11.' 1937. Alcaniz: CNT, Rioja y Navassa.
Prats, Alardo. 1938. *Vanguardia y Retaguardia: La Guerra y la Revolución en las Comarcas Aragonesas.* Santiago: Ediciones Yungque.
'Programme de la Fédération des Collectifs d'Aragon.' 1937 (1970). In Daniel Guérin (ed.), *Ni Dieu, Ni Maitre.* Paris: Maspero.
'Revolutionary Culture in Spain.' 1939. *Spain and the World, 2* (45-6) (3 Dec.), 3.
'Revolutionary Economy at Valjunquera.' 1936. *Spain and the World, 1* (2) (24 Dec.), 3.
'Revolutionary Economy: Collectivisation in Graus.' 1937. *Spain and the World, 1* (20) (22 Sept.), 3.
'Revolutionary Economy: the Wood Industry of Cuenca.' 1938 *Spain and the World, 2* (18 Feb.), 3.
Semo, E. 1973. *Historia del Capitalismo en Mexico, Los Orgenes 1521-1763.* Mexico: D.F. Editoral.
'Social Aspirations and Achievements of the Spanish Peasant.' 1936. *Spain and the World, 1* (2) (Dec.), 2-3.
Souchy, Augustin. 1937a. *Collectivisations: L'oeuvre constructive de la révolution espagnole.* Toulouse.
—, 1937b. *Entre los campesinos de Aragón,* Barcelona: Ediciones Tierra y Libertad.
—, 1938. 'The Economic Council of the Spanish Workers,' *Spain and the World, 2* (30) (18 Feb.), 3.
Tambaret, R. 1938. 'Peasant Collectivity of Balsareny,' *Spain and the World, 2* (27) (5 Jan.), 2.
Thomas, Hugh. 1961. *The Spanish Civil War.* New York: Harper Colophon.

Tierra y Libertad. 1936-9. Barcelona. FAI. 'UGT & CNT Collectives: A Comparison,' 1938, *Spain and the World, 2* (44) (12 Nov.), 3.
Vicens Vices, Jaime. 1969. *An Economic History of Spain.* New Jersey: Princeton.

Part Two

HORIZONS OF INQUIRY

5 HUMAN GEOGRAPHY AS TEXT INTERPRETATION

Courtice Rose

> As soon as we agree that the purpose of every science is accomplished when the laws which govern its phenomena are discovered, we must admit that the subject of geography is distributed among a great number of sciences; if however, we would maintain its independence, we must prove that there exists another object for science besides the deduction of laws from phenomena. And it is our opinion that there *is* another object – the thorough understanding of phenomena. Thus we find that the contest between geographers and their adversaries is identical with the old controversy between historical and physical methods. One party claims that the ideal aim of science ought to be the discovery of general laws; the other maintains that it is the investigation of the phenomena themselves – Franz Boas, 1887.

It is only a little ironic that the 'physicist/cosmographer' distinction drawn so aptly by Franz Boas almost a century ago is still very much a part of the continuing debate into the epistemological quagmire surrounding the practice of human geography. Boas was quite clear about the difference of opinion between 'physicists' and 'cosmographers:' 'While physical science arises from the logical and aesthetical demands of the human mind, cosmography has its source in the personal feeling of man towards the world, towards the phenomena surrounding him' (Boas, 1887, 139).

That we are not now clear as to which side of the naturalist-historicist continuum human geography belongs, or for that matter, if it can be meaningfully treated as emanating from either perspective, is convincing testimony to the fact that the question of human geography as one of chorology, as spatial science, as behavioral science, or as human ecology, is a question that remains quite unresolved. Despite recent attempts by geographers to ally themselves with (1) a scientific and positivist rationale (e.g. Schaefer, 1953; Bunge, 1966; Chorley and Haggett, 1967; Harvey, 1969); (2) a perception orientation derivative of psychology (e.g. Saarinen, 1969; Downs, 1970); (3) a phenomenological or humanistic orientation (e.g. Relph, 1970; Tuan, 1971; Buttimer, 1974; Relph, 1976; Seamon, 1979); or (4) a historical/idealist tradition (e.g. Sauer, 1925; Lukerman, 1964; Harris, 1971; Guelke, 1974), there seems little or no agreement on an epistemological stance which will encompass the largely physical background of nineteenth-century geography, satisfy the positivist's demands for the explanation of events in human geography by the derivation of laws and theoretical principles, and still serve as a framework for more recent phenomenological investigation of such topics as 'at-homeness,' 'dwelling,' 'encounter,' and 'sense of place' (Entrikin, 1977). A number of questions are thereby posed: do human and physical geographers study the same kind of phenomena? Do the premises of human geography as a separate discipline have any independent theoretical origin? What is the status of 'unobservable' events in geography? Is the goal of human geography to explain events or to understand events in a wider context? In short, what exactly is it that geographers *do* when they practice human geography?

This essay suggests that doing human geography consists of interpreting *texts* – an activity much like that of ordinary reading. A text can be considered as any set of signs, symbols, verbal or non-verbal gestures and actions housed in a pattern of linguistic meanings. The texts of human geography, I argue, comprise not only words, numbers and geometric symbols such as are found in the usual definition of texts as any piece of written communication, but also the whole gamut of expressive signs, whether these signs are found on maps, in the field, in equations, in spoken or written accounts, or as acted out in the events which the geographer is attempting to comprehend. Further, I suggest that the activity of human geography is the interpretation of such texts: the apprehension of meanings openly expressed by the signs and symbols as well as those meanings which are only implicit in the text taken as a whole.

Following a framework suggested by Beck (1975), I argue that human geographers might pursue their study in a context which demands that they be *both* actors in and spectators of the social world into which they inquire. As such, they are required to participate in two different kinds of mental events: (1) 'seeing,' in the sense that any agent sees the sense of his thoughts and actions – i.e. he sees them to be meaningful to him in a commonsense fashion, and (2) 'seeing,' in the sense that an observer or spectator perceives an action and is able to interpret it correctly – i.e. he is able to give an account of it, either scientifically or otherwise. While acting as a spectator of a particular event he can interpret optical data as possible clues to what is going on in the actor's mind but obviously he cannot bring the same conception to bear on the event as the agent himself can.

'Seeing' as an event for the actor is thus not the same as 'seeing' as an event for the spectator. In the case of map-making, for example, both the actor (the map-maker) and the spectator (the human geographer) see the same optical data; they see a person making marks on a piece of paper with a drawing instrument. In other words, they both see 'paper-marking' taking place. But it would also be true that the agent of the action involved could state that 'map-making' is actually taking place without having to examine his 'optical data' in any deliberate manner. Such is not the case for the human geographer as spectator: he must come to interpret his set of optical data by interpreting it according to rules – e.g. for the purposes of explaining or understanding the event of map-making. Why is the 'seeing' as an event for the actor not the same as the 'seeing' as an event for the spectator? And further, how does the human geographer as spectator come to agree with the human geographer as actor as to the meaning of this particular act?

Observation and Interpretation

A first set of questions relates to observation and interpretation. In what sense can the actor and the spectator be said to see the same thing and yet not see the same thing? One common formulation would be that since observation is a 'theory-laden' undertaking, the observation of 'paper-marking' is shaped by prior knowledge of 'paper-marking.' The spectator thus sees 'paper-marking' in a referential use of the word 'seeing' – i.e. he sees what is referred to by the linguistic definition of 'paper-marking' as a subcategory of all possible human actions. The actor, on the other hand, sees 'mapping' – i.e. he sees the sense which this 'paper-marking' has for him in his objective of making a map. Accordingly, the argument runs, if seeing is theory-laden in the sense described, then two proponents of different theories cannot strictly observe the same things (Hanson, 1958). Thus the account or report given by the spectator would be just as 'objective' as the account given by the actor, since the two theory-laden observations derive from *different* sets of categories, not because they have the same sets of categories but simply different ways of organizing the items in those categories.

One way out of this dilemma might be to hold that in any real 'seeing,' there is some 'brute' data or 'given' which is embedded in the actual experience of observation and remains constant and unaltered. The 'given' then is that which should arbitrate between two different accounts of 'seeing' the same thing. Thus it would be held that there is some observable clue in the seeing of 'paper-marking' which should make it obvious to the spectator that the paper-marking stands for or indicates that this activity is really a subcategory of a more general activity called 'mapping.'

Without even referring to some of the psychological evidence in this matter, common sense would indicate that there can be no such immunity to conceptual change – beliefs, mind sets, expectations and, more importantly, language itself exercise an enormous role in determining the quality of anything which is sensuously given. If the 'given' were embedded in any particular concept, such as 'mapping,' and if a language were available to exactly describe that 'given-ness,' there would presumably be no need for the actor to differentiate between 'paper-marking' and 'mapping' as separate but related descriptions of what the spectator had seen – they would automatically be 'mapping' for the spectator. The fact that they are not indicates that while for the spectator such 'paper-marking' is merely that, for the actor the 'paper-marking,' when it is pointed out as such by the spectator, is more like a set of hypotheses such as the following:

C_1 People acting so as to make sets of marks such as these on paper are usually engaged in mapping.
C_2 This person is making such marks on paper.
Therefore, this person is engaged in 'mapping.'

Now, since 'paper-marking' can also be interpreted as writing, drawing, sketching, making errors, designing, printing, etc., hypotheses about 'paper-marking' can obviously be mistaken and thus the form of such an argument is always probabilistic. On the other hand, the categorizations of the spectator cannot so easily be subjected to statements about their certainty. Concepts such as 'paper-marking' in and of themselves are not capable of making any truth claims. Or to think of it in another way, the 'seeing' of the spectator is akin to the words of a language in the sense of being a set of signs (or phonemes) in a prearranged vocabulary; on the other hand, the 'reporting' of these signs is clearly a different operation and represents a body of assertions (or hypotheses) expressible by means of that vocabulary. This is the case only because the spectator (or more properly, the reporter) can put such a series of signs together into a simple association such as: 'here is mapping' or 'these people are making maps.' Only when language is taken in this second sense – i.e. language as a vehicle for reporting worldly phenomena – is it capable of being evaluated with respect to a truth claim. It is, therefore, only at the level of a sentence which at one and the same time refers to hypothesized activity ('mapping'), and which also says something about such an activity (i.e. it is 'conventional mapping consisting of making certain marks on paper'), that it is possible for the spectator to identify with some degree of certainty what he is actually seeing.

Moreover, there can exist considerable ambiguity in such sets of hypothetical sentences and it is exactly within this ambiguity that the problem of meaning arises for the spectator. On the one hand, the actor 'sees' what the spectator sees – i.e. they both see 'paper-marking,' but the spectator reports such an event to himself differently, since he not only categorizes but hypothesizes about the contents of this categorization.[1] Crucial to this problem of ambiguity is that even given the same categorization as the actor it is possible for the spectator to have alternative assignments derivative of different hypotheses. The fact that the actor can inform the spectator that 'paper-marking' in cartography laboratories *can* be called mapping does not destroy the spectator's previous set of categories about what counts as 'paper-marking' – these features merely remain as sets of conflicting or complementary hypotheses for him.

The upshot of this first set of questions is that even if observation is construed as fully categorized, hypotheses about those categories can still be totally ambiguous in character. Second, that if we are to begin to make any sense of the nature of this ambiguity, we must drop the basic premise that the 'given' or some type of 'brute' data alone is sufficient to differentiate between conflicting hypotheses. In short, to get at the meanings inherent in different actor situations, we must consider that the language itself is partly constitutive of the situation we study. There is in this sense something very artificial about a reality which is construed as 'something given' *along with* a separate language in which to describe that 'given.' More likely it is that the 'given' (if it exists) and its terminology are continuously bound up together.

'Seeing' and Meaning

Our disgruntled spectator might then question the actor's descriptions of 'mapping' from a different tack. He might rightly decide that the actor's language itself has something peculiar about it and say,

> If we agree that my categories are not the problem, perhaps it is the meanings which I attach to those categories which makes the difference. Let us simply decide which meanings we have in common and then surely we can agree to what constitutes 'mapping.'

We thus have a second set of questions concerning the relationship of 'seeing' and meaning. One formulation might be that alternative hypotheses about actions, perceptions, and events can confer alternative meanings on the categorization systems in question. The meaning of a category is therefore dependent on the language in which it has a place. Change the language and one automatically changes the meanings of his or her categorical referents. In this regard, the spectator must come to share the language of the actor in order to see the 'paper-marking' as constitutive of mapping.

There is no neutral ground here. To understand how I 'see,' you must understand how I mean something in *my* language; conversation requires conversion. However, meaning as the thing 'referred to' is quite different than meaning as 'the idea expressed,' 'the sense,' or 'the gist.' The former is a matter of convention derivative of language seen as a set of more or less flexible categories – categories which are directly derivative of use – e.g. in cartography laboratories, 'mapping' means 'making marks on paper.' In the latter sense, meaning as 'gist' or 'thrust' or 'idea expressed' is not an appendage-like characteristic of a term, but rather inheres *in* a term and implies the performance of *other* actions and events as well as those directly in question. But if few people would disagree with the notion that meanings are wholly relative to language systems, we would not then wish to make the further statement that meanings cannot be shared across different language systems.

Therefore, terms may denote very same things (identity of reference) even if their relations to other events, things, and ideas are catalogued quite differently (an identity of sense). Thus for the spectator, 'paper-marking' and for the actor 'mapping' refer to the same phenomenon. What these terms signify to each observer – i.e. what he *takes* these signs to be for him – is irrelevant as long as some synonomous relation can be set up between what counts as 'mapping' for the actor and what counts as 'paper-marking' for the spectator. Thus referential meaning does not wholly depend on some physical or observable constitution of terms and it is therefore very possible to share common meanings with those who would disagree with us about our sets of referential meanings.

But what of the other connotation of meaning, meaning taken as 'sense-giving.' Is it possible to have the same shared meanings in the sense of 'expressing the same ideas as' even when we begin from different belief systems? What I would propose here is that the locus of such meaning-giving is not to be found at the level of the word or of the sentence but at the level of the text. Texts, when seen not simply as sets of signs, symbols and gestures either spoken or written out, but when taken as purveyors of a pattern of actions and events which is not present, or observable, but which is implicit in the 'gist' or 'thrust' of the event of which the text speaks as a whole – this function of text is what 'positions' or gives direction to a set of statements made about the sense of a particular phenomena.

Taken in this fashion our example of the term 'mapping' would obtain its referential meaning from the word or sentence meanings already present in the lexicon or vocabulary. In other words, this particular set of phonemes 'm-a-p' refers to pieces of paper marked in a specific fashion, or at the sentence level 'mapping' means the activity of putting certain types of symbols and signs on paper. At that level which concerns only the association of such signs with certain worldly phenomena, a map is a 'text' which has agreed-upon meanings for geographers, engineers, planners, airline pilots, travelling motorists, and numerous other map-users. Thus the spectator would simply function at this level by adhering to a set of rules which the actor could supply him as to what counts as 'mapping.' Human geography taken in this sense is nothing more (and nothing less) than rule-following.

Texts and Implied Events

However, quite a different text and interpretation is required of the spectator (the human geographer) when he is asked to understand implied events. Suppose that the spectator observes the actor in question peering closely at his map, looking away for a moment, then scratching out a mark on the paper, then finally throwing down the pen and leaving the room. Here there is no first-order connection which can be made between these events and how the spectator is to interpret them, but there is a text available. By his reflection about these perceived events, the spectator can create for himself a text or a set of related thoughts and ideas attendant upon the actions implied by what he has seen.

In this text 'event,' the signs, symbols, thoughts, and feelings do not originate in the perceived external events but are wholly internal to the spectator and they go beyond the events themselves toward the pattern that these events imply as present in a larger nexus of meanings for the actor. What is presented to the spectator in this text event is not what is optically seen, but rather what is implied as the missing parts of a coherent whole.[2] The spectator cannot directly experience the frustration of the actor who has just made a mistake and become angry, but he can 'see' that this series of events is coherent with the whole experience known as 'being angry at oneself.' There is a pattern to the actions of the actor which the spectator grasps by seeing only a portion of that pattern; the parts lead toward or disclose that whole. More importantly, the spectator can understand the meaning of this whole, having actually seen only one portion of it, because he knows what it means to be in a position similar to that of the actor. Thus the text here refers to a series of implied events *in a pattern for the actor,* which the spectator because he is also an actor would expect to occur – e.g. other events typical of angered individuals such as shouting at others or slamming doors. But it is the text taken as a whole which delivers this pattern for the spectator. Just as it is not the sentence which gives the gist of a story, the thrust of a speech or the meaning of a poem, it is the force and direction of the text as a whole which positions the event in a meaning pattern for the spectator.[3]

Sense meanings for spectators can thus be constituted through text events which are 'spectator-as-actor' derived; they primarily indicate the reasons, motives, intentions, dispositions, and emotions which are not immediately apparent to the spectator but which, upon reflection, are implied as initially credible for the actor involved. Thus the meaning of a particular event is disclosed to the spectator when taken in the 'spectator-as-actor' text.

The importance of this initial credibility characteristic of text interpretation is that it allows the spectator to make statements about particular actors' reasons for acting because he is also a member of a set of actors. Both actor and spectator must therefore know what it means to 'become angry;' hence for the spectator to cite the 'correct' reason is for him to give the same reason that the actor would for the particular event in question. And although it is clearly the case that the imputed or assessed reason given by the spectator may not be the exact reason the actor had for doing what he did, a *prime facie* claim for credit is allowed for the actor by the spectator, since if there were other reasons involved then presumably the action or the reasons adduced by the spectator would have been different.[4]

There is always a relationship of appropriateness or 'fitting' between reasons and their accompanying actions – this fitting together of reasons and actions is made apparent in the pattern available from the text.

Ultimately the spectator makes sense of the event in question by performing a thought experiment in which he is able to see the *practicability* of the reasons that he himself understands could be the case for the actor. I am thus suggesting that reasons (unlike causes) can be shared between an actor and a spectator and the medium of this sharing can be the interpretation of a particular text which the spectator reads to the best of his ability.

A corollary of this notion of sense meanings as derivative of text interpretation is that the understanding of reasons cannot be disconfirmed by empirical evidence in the same manner as can causal or scientific explanations. Reasons can be appropriate or inappropriate, adequate or inadequate, suitable or unsuitable, practical or impractical, but never true or false. As spectators, then, we never have any evidence against the *having* of reasons but only evidence against *not* having such reasons due to the fact that they are not adequate, appropriate, or practical for the pattern evident at the level of the text. We thus do not expect the frustrated map-maker to reappear whistling or singing cheerfully or looking at his map in a very approving manner, and saying, 'This map is a piece of fine work.'

The upshot of these latter deliberations is that (1) spectators can understand the actions of actors through the apprehension of patterns in texts derived from their own thought experiments; (2) that the possibility of a spectator citing the same reason for an action as the actor rests both on the part-whole relationships within such patterns and on the fact that spectators are both actors and spectators of actors; hence (3) statements that spectators make about the reasons that actors have cannot be epistemologically prior to those made by the actors involved.

Conclusion

Within this view of human geography as text interpretation, there still exists the dilemma which faced the nineteenth-century 'cosmographers.' If we accept the premises that: (1) human geography takes place entirely within a sphere of signs; (2) this sphere of signs must have intersubjective origins; and (3) it is the job of the human geographer to stand 'outside' this sphere of signs as a spectator in order to make certain statements about actors operating within the sphere of signs and, further, to communicate the meanings of such phenomena as he should deem important back to the actors involved, then we cannot escape the conclusion that the assumption of a sign system in which perception, speaking and reporting have as much bearing on the reported phenomena under study as the phenomena itself.

Thus perception, when considered as an activity 'in' a reflexive symbolic system, cannot be easily divorced from its product. What is perceived is not a sign or symbol detached from a specific meaning or interpretation but rather a sign as part of a text-event, a pattern in which the sign or symbol and its interpretation are often inseparable for the actor. The 'doing' of the action, the 'saying' of the symbol and the 'knowing' of the meaning, in other words, are all very much overlapped. Hence geographers who take text interpretation as defining what they 'do' when they operate as spectators of the human action must not strictly observe the *exact action* to which the signs, words, and sentences refer but rather must attempt to rediscover the meaning of these actions from a text – a text which is already part of the actor's world *and* the geographer's own.

Much of the existing work in human geography has tended to support tacitly a theory of evidence in which it was assumed that what would count as legitimate subject-matter for the discipline is either physically observable or able to be inferred from observable behaviour, events, and actions. All other 'evidence' has been taken as private, subjective, and therefore epistemically inaccessible. The argument presented here demonstrates that in making such an assumption human geographers have often assumed away the very linguistic preconditions for any knowledge aimed at understanding the fullness of human action. It was the social world as already constituted in actor language that was the basis of 'cosmography' for Boas, and it is this same social world given through text-events that now provides the basis for contemporary human geography. Any epistemic claims which do not recognize this fact will remain restricted either to attempts which either 'objectify' the subjective portion of human geography or assume that the subjective is not available for inspection at all.

Notes

1. Or in Wittgenstein's terms, 'seeing x' in contrast to 'seeing x as something' (Wittgenstein, 1953, 193-208).

2. The notion that meaning is related to part-whole relationships has been a powerful idea in the study of human perception. In Kant it appears as the uniting of external representations with internal consciousness in the doctrine of 'transcental apperception' (Kant, 1965, s. A105-10, s. 222-5). Dilthey's argument from 'analogy' speaks of part-whole relations as fundamental to the higher forms of understanding, i.e. understanding the 'life-expressions of others.' (Dilthey, 1927,

189-200). In Husserl's idea of 'appresentation,' the ego is constantly synthesizing various possible images of an externally perceived object, each one of which is indicative of the whole object (Husserl, 1913, s. 44, 1931, s. 49-53 and 1938, s. 18-21). Henri Bortoft also describes meaning as 'hologrammatical' in the sense that we grasp the whole through perceiving only a portion of it: 'hologram' = 'writes the whole' (Bortoft, 1971); and Beck takes all physiognomic perception to be 'synecdochic' – i.e. seeing the 'whole' because we see a portion of it (Beck, 1975).

3. What I wish to suggest here is the 'hierarchical' notion of interpretation. The hierarchy is taken from the progressive complexity found in linguistic systems with phonemes, single signs or symbols as the 'lowest' order, followed by words, phrases, sentences and finally whole texts. As one progresses up the hierarchy, there are two contrasting features apparent: (i) an increasing specificity of meaning with regard to a particular usage; (ii) a decreasing temporality associated with the sign or symbol.

4. Unless, of course, the actions involved were either involuntary in nature or deliberately made to deceive the spectator in some way. In these cases, other factors would prevent the spectator from understanding the actions of the actor in the sense implied here (see Austin, 1946; Malcolm, 1958).

References

Austin, J.L. 1946. 'Other Minds,' *Proceedings of the Aristotelian Society,* Supp., *20,* 148-87.

Beck, L.W. 1975. *The Actor and the Spectator.* New Haven, Conn.: Yale University Press.

Boas, F. 1887. 'The Study of Geography,' *Science, 9* (210), 137-41.

Bortoft, H. 1971. 'The Whole: Counterfeit and Authentic,' *Systematics, 9* (2), 43-73.

Bunge, W. 1966. *Theoretical Geography.* Series C, n.1, Lund, Sweden: C.W.K. Gleerup.

Burton, I. and Kates, R.W. 1964. 'The Perception of Natural Hazards in Resource Management,' *Natural Resources Journal, 3* (3), 412-41.

Buttimer, A. 1974. *Values in Geography.* Washington, DC: Association of American Geographers, Resource Paper No. 24.

Chorley, R.J. and P. Haggett, 1967. *Models in Geography.* London: Methuen.

Dilthey, W. 1924. *Gesammelte Schriften.* V. VII, ed. B. Groethuysen. Stuttgart: B. G. Tuebner.

Downs, R. 1970. 'Geographic Space Perception: Past Approaches and Future Prospects.' In C. Board (ed.), *Progress in Geography,* pp. 65-108. London: E. Arnold.

Entrikin, J.N. 1976. 'Contemporary Humanism in Geography,' *Annals of the Association of American Geographers, 66* (44), 615-32.

Guelke, L. 1974. 'The Idealist Alternative in Human Geography,' *Annals of the Association of American Geographers, 64* (2), 193-202.

Haggett, P. 1965. *Locational Analysis in Human Geography.* London: E. Arnold.

Hansen, N.R. 1958. *Patterns of Discovery.* Cambridge: Cambridge University Press.

Harris, C. 1971. 'Theory and Synthesis in Historical Geography,' *Canadian Geographer, 15* (3), 157-72.

Harvey, D. 1969. *Explanation in Geography.* London: E. Arnold.

Husserl, E. 1913. *Ideas: General Introduction to pure Phenomenology,* trans. W.R. Boyce Gibson. New York: Humanities, 1931.

—, 1931. *Cartesian Meditations,* trans. D. Cairns. The Hague: M. Nijhoff, 1960.

—, 1938. *Experience and Judgment,* ed. L. Landgrebe, trans. J.S. Churchill and K. Ameriks. Chicago: Northwestern University Press, 1973.
Kant, I. 1965. *Critique of Pure Reason,* trans. N.K. Smith. New York: St Martin's.
Lowenthal, D. 1961. 'Geography, Experience and Imagination: Towards a Geographical Epistemology,' *Annals of the Association of American Geographers, 51* 241-60.
Lukerman, F. 1964. 'Geography as a Formal Intellectual Discipline and the Way in which it contributes to Human Knowledge,' *Canadian Geographer, 8* (4), 167-72.
Malcolm, N. 1958. 'Knowledge of Other Minds,' *Journal of Philosophy, 55,* 969-78.
Relph, E. 1970. 'An Inquiry into the Relations between Phenomenology and Geography,' *Canadian Geographer, 14* (3), 193-201.
—, 1976. *Place and Placelessness.* London: Pion.
Saarinen, T. 1969. *Perception of Environment.* Washington, DC: Association of American Geographers, Resource Paper No. 5.
Sauer, C.O. 1925. 'The Morphology of Landscape.' In J. Leighly, *Land and Life: A Selection of the Writings of Carl Ortwin Sauer.* Berkeley: University of California Press, 1963.
Schaefer, F. 1953. 'Exceptionalism in Geography: A Methodological Examination,' *Annals of the Association of American Geographers, 43,* 226-49.
Seamon, D. 1979. *A Geography of the Lifeworld: Movement, Rest and Encounter.* New York: St Martin's Press.
Tuan, Yi-Fu. 1971. 'Geography, Phenomenology and the Study of Human Nature,' *Canadian Geographer, 15* (3), 181-92.
—, 1976. 'Humanistic Geography,' *Annals of the Association of American Geographers, 66* (2), 266-76.
Wittgenstein, L. 1953. *Philosophical Investigations.* New York: Macmillan.

6 SOCIAL SPACE AND SYMBOLIC INTERACTION

Bobby M. Wilson

The purpose of this chapter is to interpret the concept of social space within the context of Mead's (1934) theory of symbolic interaction. Past studies of socio-spatial patterns have generally failed to provide the necessary linkage between the objective and subjective components of social space. Any attempt, however, to provide this linkage will demand further insight into human and spatial interactions. A framework for such an insight can be found at least partially in the theory of symbolic interaction.

Since it was first introduced in the 1890s by Emile Durkheim (1933), the notion of social space has had a very ambiguous development. In North America, the lack of communication between disciplines probably has contributed more than any other factor to the ambiguity that exists concerning certain dimensions of the social space concept. It seems ironic that the notion, as geographers have accepted it, developed through communication between Sorre, a geographer, and Chombart de Lauwe, a sociologist-ethnologist. Sorre, relying a great deal on the work of Durkheim and impressed by the Chicago School of human ecology, suggested that social space had more than a subjective and sociological dimension – that is, it also had an objective or concrete dimension for which traditional geographic indices, e.g. land use, habitat, etc., could provide useful surrogates (Buttimer, 1971, 130).

Sorre saw the need to integrate the subjective and objective dimensions in the development of the social space concept. On a global scale he envisioned

> social space as a mosaic of areas, each homogenous in terms of the space perceptions of its inhabitants. Within each of the areas a network of points and lines radiating from certain *points privilégiés* (theatres, schools, churches, and other foci of social movement) could be identified. Each group tended to have its own special social space, which reflected its particular values, preferences, and aspirations. The density of social space reflected the complementarity, and consequently the degree of interaction, between groups (Buttimer, 1969, 419).

Notions similar to the social space concept have appeared, especially in America, but generally they have lacked the integrative power that Sorre envisioned. As a result, studies of social space often adopt separate and distinct levels of approach, focusing either on objective or subjective points of view.

The Objective Component of Social Space

Those researchers who approached the study of social space from an objective point of view tended to emphasize spatial patterns of behavior conditioned by ecological, cultural, and socioeconomic factors. As a reaction against the environmentalism of Ratzel's 'Anthropogeographie' (1882 and 1891), Durkheim emphasized the sociological dimension of differentiation in space. In America, Sorokin was one of the first social scientists to define an objective social space within a purely sociological framework. Sorokin's definition did not specify anything about one's position in geodesic space, but rather emphasized the 'horizontal dimension which ranked one's membership group in relation to other groups within the social universe, and the vertical dimension which defined one's status and role within each group' (1928, 7).

Unlike Sorokin, social area analysis did attempt to incorporate the spatial component of social space. Beginning with studies of Los Angeles and San Francisco done by Shevky and William (1949), and Bell (1953), this research grouped census tracts according to three major indices – social rank, urbanization, and segregation – thought to describe the way in which urban populations are differentiated in space. This approach was later adopted by geographers in factorial ecology studies (Murdie, 1969; Berry and Horton, 1970). Unlike the social area studies, however, which defined space bounded by boundaries of census tracts, factorial ecology studies were concerned with the spatial patterning of census-tract groupings. These studies indicated that social rank, urbanization, or stage of the family cycle, and segregation were organized in sectoral, concentric, and clustered patterns. The utilization of computer technology within the social sciences provided the necessary tool for large-scale analysis in order to discern such patterns.

As a reaction against macro-scale studies of socio-spatial patterns, some researchers began to study action spaces, activity spaces, behavioral fields, and similar concepts, which attempted to represent a more dynamic variation of social space at the micro-level (Chapin and Hightower, 1966; Horton and Reynolds, 1971; Cox and Golledge, 1969)

Within this approach, the nature of people's interaction in space is taken as a critical variable to the way they experience space. The role of social interaction as a critical clue to people's movement in space can be described, for example, by the study of racial discrimination, which leads to the spatial segregation of residence and other activities (Wheeler, 1971). As a result, a geographically closed system of social interaction may operate. However, within a closed system – e.g. a black ghetto – there may be a number of semi-closed systems among which certain social systems may play a prominent role. Dominant social systems such as church or clubs reflect degrees of social stratification which may foster a further areal concentration of social interaction. Certain patterns of activity may thereby occur which can be explained in terms of action spaces, activity spaces, and behavioral fields. Insights into the nature of activity patterns have often been complemented by literature on social interaction in urban space, which is found to vary according to life-styles, stage in the family cycle, and social status (Michelson, 1970; Gans, 1962; Young and Willmott, 1957).

The Subjective Component of Social Space

The failure of many geographers to deal with the human side of the equation in studying spatial patterns in the past has led many researchers to emphasize the 'city of the mind' in explaining overt spatial patterns within urban areas (Carr, 1970). The subjective component of space has mainly been studied within the framework of urban images, cognitive maps, and life space. Lynch (1960) attempted to examine the influence of the physical environment on the attributes of identity, structure, and meaning in the mental image. He defined imageability as that quality in a physical object which may have a high probability of evoking a strong image in an individual. Whereas Lynch tended to emphasize the role of the physical environment in fostering imageability, the sociologist Strauss (1961) realized that 'certain symbolic means must be resorted to in order to perceive the environment.' Taking Strauss and Lynch together, it becomes clear that the perception of the environment encompasses both psychological and physical factors.

The importance of both the physical and non-physical environment led Lewin (1936) to derive his concept of 'life space' – the totality of all possible events which may influence one's behavior. This represents an attempt to explain one's behavior according to his position among physical objects and his relationship to other persons. In representing the life space Lewin included social facts and relationships only in so far as they influenced the individual, that is, only when they are immediately present. As regards to membership in a group, he considers more the belief of the person and the way it affects him than the sociologically defined criteria of group membership (1936, 25). These limitations are a major downfall of Lewin's life space in terms of its effectiveness in conceptualizing socio-spatial behavior.

Following the same basic idea, Lee (1968) proposed the concept of *schema* to conceptualize the mental representation of socio-physical space and argued that it was a function also of both physical environment and social interaction within a territorial base. Unlike Lewin, Lee is more concerned with social facts and relationships. Lee attempts to measure the degree of agreement and interdependence in the formation of the schema 'among any given aggregate of people selected on some independent practical or theoretical criterion' (1968, 262). He focuses mainly on such criteria as length of residence, social class, and age.

Another approach to environmental behavior has been demonstrated through the use of cognitive maps, which are forms of mental representation through which data of environmental perception are stored. These mental schemata provide one with 'a coping mechanism through which the individual answers two basic questions quickly and efficiently: (1) where certain valued things are; and (2) how to get where they are from where he is' (Downs and Stea, 1973, 10). For many researches, the interest in environmental cognition rests on the assumption that we can better understand people's behavior, needs, and desires with regard to the environment if we know how they portray it cognitively. How we know the environment, however, does not have to be based upon or linked with immediate behavior or objects. Environment can be known through the more abstract symbols of art, education, and religion.

A common fault, therefore, of the approaches analyzing the subjective component of environmental behavior is that they fail to provide a framework for understanding the process by which shared meaning within groups develops and thereby contributes to patterns of socio-spatial behavior. Although both Lee's schema and Lewin's life space tend to be socio-spatial in nature, they provide us with little conceptual insight into this process. A complete analysis of the social space concept would require such an understanding.

Symbolic Interaction

Symbolic interaction, a theory originally arising out of sociology and social psychology, has recently reappeared as a pragmatic perspective in the study of human behavior (Huber, 1973). The epistemology of symbolic interaction is derived basically from the pragmatic model of Dewey, and especially Mead, who, above all others, laid the foundation. Mead adopted the premise that human actions were in part based on things that have shared meanings – i.e. intersubjectivity. Thus, the meaning of things located in space is derived in large part from social interaction.

According to Mead's theory, variations in the spatial behavior of individuals could be seen as a function of their interaction as members within specific groups – i.e. *reference groups.* If one accepts this premise, he must then ask if reference groups are indeed influential in dictating people's spatial patterns of behavior. Work in perception has shown that overtly expressed spatial patterns may have different meanings for different societies (e.g. Burton and Kates, 1964). Also, it has been shown on a micro-level that spatial patterns and the schemata responsible for their formation vary according to membership in particular groups (e.g. Buttimer, 1972; Shibutani, 1955). At the same time, the role of any group in one's lived experiences will influence that person's behavior in space.

Blumer (1969, 149) asserts that one cannot 'recognize human beings as they are unless we interpret their actions through the situations which confront them.' Shibutani (1955) has identified three distinct group referents: (1) groups which serve as a dimension for comparing and ranking oneself in society; (2) groups to which men aspire; and (3) groups whose perspectives are assumed by the actors in an event. How one defines the situation that confronts him depends upon his organized perspective. Realizing the importance of one's organized perspective, Shibutani defines reference groups as 'groups whose perspectives are used by the actors as the frame of reference in the organization of their perceptual field.' By 'perspective' Shibutani means:

> An ordered view of one's world – what is taken for granted about the attributes of various objects, events, and human nature. It is an order of things remembered and expected as well as things actually perceived; an organized conception of what is plausible and what is possible, it constitutes the matrix through which one perceives his environment (1955, 564).

The groups of greatest importance for most people are those in which they participate directly – i.e. *membership groups* (Goffman, 1967). This emphasis on group interaction and meaning is consistent with Mead's theory of symbolic interaction, which describes how intragroup interaction contributes to the development and maintenance of a 'self.' In declaring that the human being has a self, Mead is saying that the individual can be the object of his own actions and that all of his actions are in part based upon a self-consciousness. He can act toward himself as he might act toward others. Practically all shared meaning that a person acquires toward others and place is acquired as a result of these acts. This shared meaning may become a major part of the 'self' – i.e. a person's organized perspective. Mead referred to this part of the self as the 'me.' The 'me' is the organized set of attitudes of others which one himself assumes. Cassirer's *Philosophy of Symbolic Form* (1953) starts from the presupposition that the nature of man, the way he organizes his feelings, desires, and thoughts which are contained in the self, is defined through human interaction. Language, art, religion, and history are just a few of the more formalized symbolic forms that man utilizes in expressing and organizing his perspectives. As individuals, however, we have the freedom and initiative to react to the attitudes of the others. Mead referred to this reactive nature of the self as the 'I.' The 'I' is the creative self who may choose and value independently of the others (Buttimer, 1974). In this regard, one main purpose of existentialist philosophy is to emphasize the creative self and thereby perhaps free man from total reliance upon the attitudes of the others.

The Self and Social Space

An individual's behavior in space and the self which expresses and organizes that behavior cannot be viewed as being static, but changing via a complex learning process. According to Mead:

> The self is something which has a development; it is not initially there, at birth, but arises in the process of social experience and activity, that is, develops in the given individual as a result of his relations to that process as a whole, and to other individuals within that process (1934, 135).

Litwak (1959) has viewed the learning process as incorporating an orderly change from one reference group to another. In order to have a meaningful relationship within each change of one's reference group, or any new situation in which one is involved, there may be a need for a re-evaluation of one's self. Our past experiences continually take on new meanings in the light of more recent events and must be constantly re-worked and re-evaluated in accordance with our present outlook, even to the point of repudiating past selves (Strauss, 1959). Not only is there a change or re-evaluation of the self, but also a change in one's socio-spatial pattern, which thereby remains consistent with the new self.

How successful one is in dealing with changing situations may depend upon the degree of congruence between the self and his pattern of spatial interaction. For example, on arrival in a new environment the in-migrant's activity space may expand very rapidly as a result of the fixation of certain patterns of travel – e.g. journey to work, shopping, going to church (Figure 6.1 A). This spatial pattern may be the result of initial information that is acquired through interaction within the primary groups of family and friends.

This spatial pattern might be termed a *place-based community* and serves the immediate needs and purposes of the newly arrived rural in-migrant (Hyland, 1970). Because of being unfamiliar with the new urban self, there is a great deal of imitation, i.e. a lack of meaningful communication on the part of the in-migrant. As a result, the activity space has no social meaning, except as an opportunity surface. This surface may coincide with the slum area of the city inhabitated mainly by earlier in-migrants. The lack of meaningful communication can cause psychological stress during this initial period. Although not explicitly named by Mead, this stage of self-development has been referred to as the *preparatory* stage (Meltzer, 1972).

The second stage of self-development can be called the *play* and is characterized by specific roles that one carries out. In play, certain rituals are conducted which represent what the supposed purpose and actions of the group are. Because of the desire for stability and security, the acceptance of the attitudes of significant others is primary. Although the significant others may consist of more than one group (e.g. kin, friends, co-workers, etc.), there is at this stage complementarity and, consequently, a high degree of interaction among these groups. The self is ritualized within a socio-spatial network of intimacy and high density (Figure 6.1 B).

Consider, for example, black rural in-migrants in the new urban environment (Wilson, 1974). Church participation may be a major part of their ritual carry-out. The high degree of complementarity among groups within the church is reflected in the fact that the black church is

Figure 6.1: Stages of Self as Manifested in Social Space

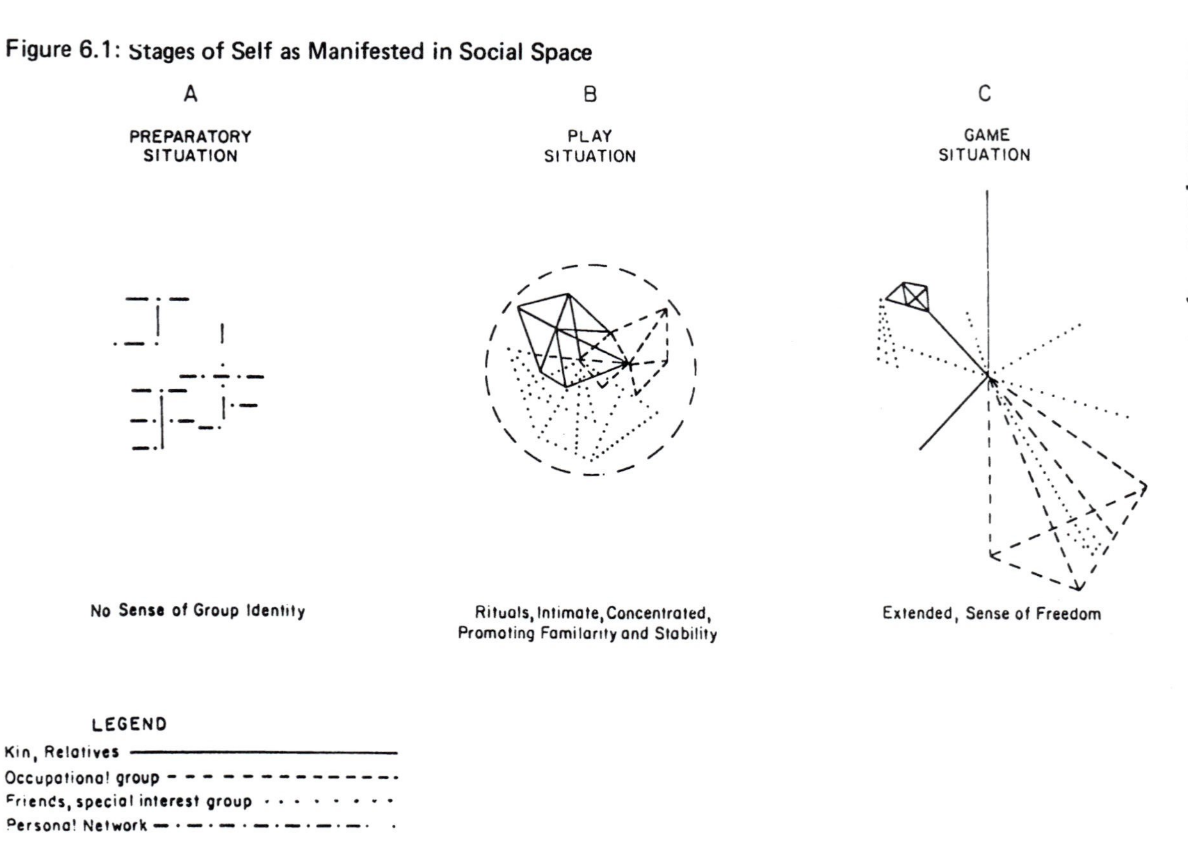

not only a place of worship but also a place of friendship and kinship interaction. The church becomes a sacred place where order is imposed upon one's pattern of socio-spatial interaction. It is independent of white control and offers a medium of intimate display of feelings. The ritualism, along with the small-scale activity pattern and well defined network, generates a social space that provides for the development and maintenance of an emotional self.

The same ritualism may be found in the small-scale geographic worlds of other lower-class, racial, and ethnic neighborhoods with their well defined socio-spatial networks. For example, the intensity of grief felt by the residents of Boston's West End relocating elsewhere is partially a function of their emotions for the area (Fried, 1969). The reference relationships of most of these residents existed in the same geographic environment. Vance Packard characterizes such a community as *authentic;* it is

> a social network of people of various kinds, ranks and ages who encounter each other on the streets, in the stores, at sports parks, at communal gatherings. A good deal of personal interaction occurs . . . all recognize it as a special place with ongoing character. It has a central core and well-understood limits. Most members base most of their daily activities in or near the community (1972, 16).

Because people are physically accessible to each other in this type of community, the means for expressing environmental experiences of the others requires little symbolic abstractedness (Wilson, 1978). A strong sense of place exists. For example, in Suttles' case study of the Adams area (1968), the locality took on a specific meaning for all its inhabitants. Because of the specific role played, abstract issues which had some significance outside the area lack significant meaning within. The play situation reinforced such spatial unity and familiarity.

Over time, however, through the person's interaction with a number of individuals and groups, his role may be no longer differentiated; his social space becomes the product of the *generalized other* – i.e. extensive reference relationships. The concept of the generalized other implies that an individual may be consistent in his thoughts and behavior even though he moves in varying socio-spatial environments (Figure 6.1C). Mead referred to this stage of self-development as the *game.* To avail himself of new situations, he often has to be able to transcend the local ritualism and restricted interaction of a place-base community. According to Thomas and Znaniecki (1974, 1856), it is the 'creative

[self] . . . that searches for new situations in order to widen the control of [one's] environment.' However, such opportunities often come about at the expense of losing the security that characterizes the self at the play stage. Because of the extensive nature of the generalized other, intragroup interaction provides for less emotion and security and greater reliance on an abstract self:

> Such communities . . . fall short of satisfying the human need for communal relationships. For one thing, an individual characteristically belongs not to one, but to a number of different communities of interest. Since relationships between members of such interest groups tend to be segmental and transitory, their members tend to avoid becoming deeply dependent on one another. Another problem stems from the fact that communities of interest often have no physical center or boundaries. Spread out all over a city or region, this sort of community lacks any sense of permanence beyond the motivations of individual members . . . One frequently does find in them. . . those for whom the major problem seems to be finding some way to order their lives, that is, to rid themselves of some of the surplus freedom with which their senses have been overwhelmed (Zablocki, 1971, 293-4).

The infrequency of spontaneous contacts and a lack of spatial proximity cause the self to develop without direct reference to others and environment. The product of such an association is a 'community without propinquity' (Webber, 1963). Instead of being set by territory and face-to-face group interaction, the boundaries of such a community are often set by the spatial limitlessness of typographic and electronic communication. The movement away from the local ritualism of a place-base community has been aided in part by the internalization of the phonetic alphabet, which translated man from the mode of oral sound to the more abstract visual mode of knowing others and environment (McLuhan, 1962; Ong, 1967). According to Tuan (1974, 10), there develops as a result of this transformation a tendency to regard seen objects as 'distant,' thus not calling forth any strong emotional or personal response toward others and place.

Because man is constantly in a state of becoming, Mead's three-stage typology of the self should not be interpreted to mean that one stage involves a higher degree of development than the others. There are groups in society who maintain continuously a place-based community. They should not be considered less developed than their more abstract and cosmopolitan counterparts at the game stage. Also, none of the stages are ever completely and absolutely realized by an individual or group. The learning process involves the incorporation of earlier stages, rather than replacement of stages. There is not an emotional self which lacks completely the creativity and abstractedness of the game stage; no abstract self that is not emotional and secure in certain respects. Even in very familiar situations, individuals often experience the feelings of alienation and stress that characterize the preparatory stage.

Finally, the self can be characterized as an ontological structure which manifests itself in social space. A place-based social space is essential for some groups' development and maintenance of self. In other groups, the self may be based on a socio-spatial interaction that is diffuse and unconnected in physical space. The whole of a person's lived experience involves to a certain extent the attempt to establish some degree of symmetry between self and external behavior in space. How successful one is in achieving this symmetrical relationship determines to an extent his success in eliminating environmental stress or alienation. In the past, these spatial experiences were to a great extent studied independently of the meaning context in which they existed. The concept of social space and the theory of symbolic interaction allow for a better interpretation and integration of social meaning and spatial experiences.

References

Bell, Wendell. 1953. 'The Social Area of the San Francisco Bay Region,' *American Sociological Review, 18,* 29-47.

Berry, Brian J.L. and Horton, Frank E. 1970. *Geographic Perspectives on Urban Systems.* Englewood Cliffs, NJ: Prentice-Hall.

Blumer, Herbert. 1969. *Symbolic Interactionism: Perspective and Method.* Englewood Cliffs, NJ: Prentice-Hall.

Burton, Ian and Kates, Robert. 1964. 'Perception of Natural Hazards in Resource Management,' *Natural Resource Journal, 3,* 412-41.

Buttimer, A. 1969. 'Social Space in Interdisciplinary Perspective,' *The Geographic Review, 59,* 417-26.

—, 1971. *Society and Milieu in the French Geographic Tradition.* Chicago: Rand McNally.

—, 1972. 'Social Space and the Planning of Residential Areas,' *Environment and Behavior, 4* (3), 217-318.

—, 1974. 'Values in Geography,' Commission on College Geography. Washington, DC: Association of American Geographers, Resource Paper No. 24.

Carr, Stephen. 1970. 'The City of the Mind.' In H. Proshansky *et al., Environment Psychology,* pp. 518-32. New York: Holt, Rinehart and Winston.

Cassirer, Ernest. 1953. *The Philosophy of Symbolic Forms,* 3 vols. New Haven: Yale University Press.

Chapin, F.S. and Hightower, H.C. 1966. *Household Activity System: A Pilot Investigation*. Chapel Hill, NC: University of North Carolina Press.
Cox, K.R. and Golledge, R.G. 1969. *Behavioral Problems in Geography: A Symposium.* Evanston, Illinois: Northwestern University Studies in Geography, No. 17.
Downs, Roger M. and Stea, David. 1973. 'Cognitive Maps and Spatial Behavior: Process and Products.' In Roger M. Downs and David Stea, *Image and Environment: Cognitive Mapping and Spatial Behavior,* pp. 8-26. Chicago: Aldine.
Durkheim, Emile. 1933. *The Division of Labor in Society,* translated and edited by George Simpson. New York: Macmillan.
Fried, Marc. 1969. 'Grieving for a Lost Home.' In Leonard J. Duhl, *The Urban Condition.* New York: Simon and Schuster, Clarion Books.
Gans, Herbert. 1962. *The Urban Villagers.* New York: The Free Press of Glencoe.
Goffman, Erving. 1967. *Interaction Ritual: Essays on Face-to-Face Behavior.* Garden City, New York: Anchor.
Horton, F.E. and Reynolds, D.R. 1971. 'The Effects of Urban Spatial Structure on Individual Behavior,' *Economic Geography, 47,* 36-48.
Huber, Joan. 1973. 'Symbolic Interaction as a Pragmatic Perspective: The Bias of Emergent Theory,' *American Sociological Review, 38,* 274-84.
Hyland, Gerard A. 1970. 'A Social Interaction Analysis of the Appalachian In-Migrant: Cincinnati S.M.S.A. Central City,' unpublished master's thesis, University of Cincinnati.
Lee, Terence. 1968. 'Urban Neighborhood As a Socio-Spatial Schema,' *Human Relations, 21,* 241-68.
Lewin, Kurt, 1936. *Principles of Topological Psychology.* New York: McGraw-Hill.
Litwak, Eugene. 1959. 'Reference Group Theory, Bureaucratic Career, and Neighborhood Primary Group Cohesion,' *Sociometry*, *22*, 72-83.
Lynch, Kevin. 1960. *The Image of the City*. Cambridge, Mass.: MIT Press.
McLuhan, Marshall. 1962. *The Gutenberg Galaxy: The Making of Typographic Man.* Toronto: University of Toronto Press.
Mead, George H. 1934. *Mind, Self, and Society.* Chicago: University of Chicago Press.
Meltzer, Bernard N. 1972. 'Mead's Social Psychology.' In Jerome G. Manis and Bernard N. Meltzer, (eds.), *Symbolic Interaction: A Reader in Social Psychology,* 2nd edition, pp. 4-22. Boston: Allyn and Bacon.
Michelson, William. 1970. *Man and His Urban Environment: A Sociological Approach*. Reading, Mass.: Addison-Wesley.
Murdie, Robert. 1969. *Factorial Ecology of Metropolitan Toronto, 1951-1961,* Department of Geography Research Paper No. 116. Chicago: University of Chicago Press.
Ong, Walter J. 1967. *The Presence of the Word: Some Prolegomena for Cultural and Religious History.* New Haven: Yale University Press.
Packard, Vance. 1972. *A Nation of Strangers.* New York: McKay.
Ratzel, Friedrich. 1882 and 1891. *Anthropogeography,* 2 vols. Stuttgart.
Shevky, Eshref and Williams, Marianne. 1949. *The Social Areas of Los Angeles: Analysis and Typology.* Berkeley and Los Angeles: University of California Press.
Shibutani, Tomatsu. 1955. 'Reference Groups as Perspectives,' *American Journal of Sociology, 60,* 562-9.
Sorokin, Pitirim A. 1928. *Social and Cultural Mobility.* London: Collier-Macmillan.
Strauss, A.A. 1959. *Mirrors and Masks.* Glencoe, Ill.: The Free Press.
Strauss, A.L. 1961. *Images of the American City.* New York: The Free Press of Glencoe.

Suttles, Gerald D. 1968. *The Social Order of the Slum: Ethnicity and Territory in the Inner City.* Chicago: University of Chicago Press.

Thomas, William T. and Znaniecki, Florian. 1974. *The Polish Peasant in Europe in America*, Vol. 2, 3rd edition. New York: Octagon Books.

Tuan, Yi-Fu. 1974. *Topophilia: A Study of Environmental Perception, Attitudes, and Values.* Englewood Cliffs, NJ.: Prentice-Hall.

Webber, Melvin M. 1963. 'Order in Diversity: Community Without Propinquity.' In Lowdon Wingo, *Cities and Space: The Future of Urban Land.* Baltimore: The Johns Hopkins University Press.

Wilson, Bobby M. 1974. 'The Influence of Church Participation on the Behavior in Space of Black Rural Migrants within Bedford-Stuyvesant: A Social Space Analysis,' unpublished PhD, Clark University, Worcester, Mass.

—, 1978. 'Knowing and Communicating Environmental Experiences,' unpublished manuscript.

Wheeler, James O. 1971. 'Social Interaction and Urban Space,' *Journal of Geography, 70,* 200-3.

Young, Michael and Willmott, Peter. 1957. *Family and Kinship in East London.* London: Routledge & Kegan Paul.

Zablocki, B. 1971. *The Joyful Community.* Baltimore: Penguin.

7 BODY-SUBJECT, TIME-SPACE ROUTINES, AND PLACE-BALLETS*

David Seamon

Phenomenology strives for the actualization of contact. As a way of study it seeks to meet the things of the world as those things are in themselves and so describe them. Geography studies the earth as the dwelling place of man. As one of its tasks, it seeks to understand how people live in relation to everyday places, spaces, and environments. A phenomenological geography borrows from both fields of knowing and directs its attention to the essential nature of man's dwelling on earth. The fact that all people are located in a world which is in part geographical is an irreducible characteristic of human existence. Be it small as an apartment or expansive as the ocean surrounding his ship at sea, as commonplace as a neighborhood or as strange as a distant country, man is housed in a geographical world whose specifics he can change but whose surrounds in some form he can in no way avoid. A phenomenological geography asks the significance of people's inescapable immersion in a geographical world. What are people as beings in a geographical world? What is the nature of human experience in the context of that world?[1]

This chapter works to demonstrate the value of phenomenological geography by exploring the phenomenon of *everyday movement in space,* by which is meant *any spatial displacement of the body or bodily part initiated by the person himself.* Walking to the mailbox, driving home, going from house to garage, reaching for scissors in a drawer – all these behaviours are examples of movement.[2]

This chapter explores the essential experiential character of movement. First, it sketches the notions of *natural attitude, lifeworld,* and *epoché* – each an essential notion in phenomenology. Second, the essay overviews the two conventional approaches to everyday movement – behaviorist and cognitive theories. Third, the essay introduces a phenomenological alternative and asks its value for environmental theory and planning.

*Portions of this chapter are abstracted from David Seamon, *A Geography of the Lifeworld* (Croom Helm, London, 1979). The author wishes to thank the publisher for permission to include these sections here.

Natural Attitude, Lifeworld, and Epoché

Phenomenology seeks the essential structures of human experience (Wilde, 1963, 20). It asks if from the variety of ways which men and women behave in and experience their everyday world there are particular patterns which transcend specific empirical contexts and point to the essential human condition – the irreducible crux of people's life-situations which remains when all 'non-essentials' – cultural context, historical era, personal idiosyncracies – are stripped bare through phenomenological procedures. Although it realizes that culture, history, and personality no doubt filter and condition patterns of living, phenomenology holds a certain given-ness to human experience which extends beyond particular person, place, or moment. The task of phenomenology is to unbury and describe this given-ness, of which people usually lose sight because of the mundaneness and taken-for-grantedness of their everyday life-situation.

In normal daily existence, people are caught up in a state of affairs which the phenomenologist calls *natural attitude* – the unnoticed and unquestioned acceptance of the things and experiences of daily living (Giorgi, 1970). The world of the natural attitude is termed *lifeworld* – the taken-for-granted pattern and context of everyday life, by which the person routinely conducts his or her day-to-day existence without having to make it constantly an object of concious attention (ibid.). Immersed in the natural attitude, people do not normally examine the lifeworld or even recognize its existence; it is concealed as a phenomenon:

> In the natural attitude we are too much absorbed by our mundane pursuits, both practical and theoretical; we are too much absorbed by our goals, purposes and designs, to pay any attention to the *way* the world presents itself to us. The acts of consciousness throughout which the world and whatever it contains become accessible to us are lived, but they remain undisclosed, unthematized, and in this sense concealed (ibid., 148).

Through a change in perspective – the *phenomenological reduction* as it is usually called – the phenomenologist makes the lifeworld a focus of attention: 'the acts which in the natural attitude are simply lived are now thematized and made topics of reflective analysis' (ibid., 148). A key tool for this reductive process is *epoché* – the suspension of belief in the experience or experienced object (Spiegelberg, 1965, 691). In performing *epoché*, the phenomenologist works to disengage himself from the lifeworld and re-examine its nature afresh. *Epoché* does not mean that the student rejects the world or his experience of it. Rather, he begins to question these things, as well as all concepts, models, and theories which attempt to describe and explain them. If he conducts *epoché* properly, he may discover that many events and patterns which he previously 'knew' become questionable, while facts that he had previously ignored or deemed insignificant emerge clearly and demand examination and description (Zeitlin, 1973, 147). Phenomenology is therefore an exploratory and descriptive discipline. First, it attempts to question radically the lifeworld and all theories designed to depict it. Second, it works to return to the lifeworld directly and describe its aspects as carefully as possible in their *own* terms.

A phenomenological geography re-examines the geographical portions of lifeworld. What, for example, is the nature of human dwelling on the earth? What experiential meanings do places have for people? How do different people experience nature and the physical environment? In what ways do people notice or fail to notice their geographical world?

The present aim is to explore everyday movement phenomenologically. Two steps are required: first, to set aside conventional theories of movement; second, to examine everyday movement as it occurs in its own lifeworld fashion.

Conventional Approaches to Everyday Movement

In conventional social science and environmental psychology, everyday movement has generally been discussed in terms of *spatial behavior* – people's movements in large-scale geographical space – and considered in two major ways. A *behaviorist approach,* linked with the philosophical tradition of empiricism, views everyday movement in terms of a stimulus-response model – i.e. a particular stimulus in the external environment (for example, the ringing of a telephone) causes a movement response in the person (the hearer gets up to answer it).[3] Attempting to imitate the methods of natural science, behaviorists have generally restricted their research to visible behaviors which can be verified through some form of empirical measurement. Strict behaviorists discount all inner experiential processes (for example, cognition, emotions, bodily intelligence) because these phenomena are seen as imprecise and only knowable to the person who reports them. Thus, the behaviorists generally study what the animal or person *does,* rather than what it, he, or she experiences. In practice, behaviorist work discussing spatial behavior as an explicit theme has generally studied rats learning to move through mazes; research with human subjects has been much less frequent.[4] Perhaps because it is best conducted in an experimental context and therefore not so easily applied to real-world contexts such as street and neighborhood, strict behaviorist theories have had only minimal impact on research in spatial behavior at the scale of geographical environment (Getis and Boots, 1971).

In contrast, theories of *spatial cognition* are associated with the philosophical tradition of rationalism and have had major impact in behavioral geography.[5] In their various forms, these theories argue that spatial behavior is dependent on such cognitive processes as thinking, figuring out, and deciding. In practice, most of this research has studied a particular individual's or group's cognitive representation of space, which is elicited by such devices as map drawings or questionnaires. The assumption is made that a study of these *cognitive maps*, as most geographers have come to call them, will lead to an understanding of the individual's and group's behavior in space. 'Underlying our definition of spatial cognition,' say Downs and Stea (1973, 9), 'is a view of behavior which, although variously expressed, can be reduced to the statement that human spatial behavior is dependent on the individual's cognitive map of the spatial environment.'

A major weakness, phenomenologically, of both the cognitive and behaviorist approaches is their insistence on explaining spatial behavior through an imposed *a priori* theory. The cognitive theorists assume that the cognitive map is the key to spatial behavior, while the behaviorists look toward the sequence of stimulus-response. In performing *epoché,* the phenomenologist breaks away from these two opposing views and returns to everyday movement as it is a phenomenon in the lifeworld. On one hand, he brackets the assumptions that movement depends on cognitive map; on the other, that movement is a process of stimulus-response.

To carry out this bracketing process practically, the phenomenologist must return to everyday movement as experience – as it happens in its own fashion with its own structure and dynamics. To do this, he has various approaches open to him. For example, he might carefully reflect on movement as he experiences it in his own life. Alternately, he might gather accounts of movements as described by others – through interviews, open-ended conversations or accounts in imaginative literature. The present phenomenology of movement makes use of a collection of descriptive reports gathered from a group of people who were interested enough in their own day-to-day contact with the geographical world to meet weekly for several months and share in group context personal experiences relating to everyday movement and other related themes.[6] Out of these *environmental experience groups* there gradually arose several characteristics of everyday movement of which three will be discussed here: (1) the habitual nature of everyday movement; (2) the importance of the body; (3) body and place 'choreographies.' Each theme is considered in turn, then contrasted with conventional behaviorist and cognitive theories.

The Habitual Nature of Everyday Movement

> When I was living home and going to school, I couldn't drive to the university directly – I had to go around one way or the other. And I once remember becoming vividly aware of the fact that I always went there by one route and back the other – I'd practically always do it. And the funny thing was that I didn't really have to tell myself to go there the one way and back the other. Something in me would do it automatically – I really didn't have much choice in the matter. Of course, there would be some days when I would have to go somewhere besides school first, and so I would take a different route, but otherwise I would go and return the same streets each time – a member of the environmental experience group.

A habit is any acquired behavior that has become more or less involuntary. One of the first characteristics of everyday movement to which the environmental experience groups brought attention is its habitual nature. Consider the above observation. The group member makes automatic use of the same route sequence in her daily driving pattern. Her path movements happen 'by themselves,' so to speak, without the necessary intervention of her conscious attention: 'Something in me would do it automatically – I really didn't have much choice in the matter.' Phrases from other group observations reflect this same self-acting quality: 'You go and you don't even know it;' 'I did the trip so effortlessly and unconsciously;' 'I always want to go the same old rote way.'[7]

Habitual movements extend over all environmental scales – from driving and walking to reaching and finger movements. One group member described how she sometimes goes to class and wonders how she got there, simply because she has no recollection of the walking experience: 'You don't remember walking there – you do it so automatically.' Another group member explained that he had recently switched rooms with his apartment mate, yet occasionally he finds himself going to the old room rather than the new: 'It doesn't register with me that I've headed for the "wrong" place.' A third group member sometimes forgot to place a clean towel under the sink after he had taken the dirty one to the laundry; when he washed dishes, he found himself reaching for the towel, even though a few minutes before he had already looked and not found it there. A fourth group member reported that in making telephone calls, he caught himself several times dialing his home phone rather than the number he had planned to call: 'My thoughts will be elsewhere and my fingers automatically dial the number they know best.'

Regardless of the particular scale at which they happen, these observations suggest that many movements are conducted by some preconscious process which guides behaviors without the person's need to be consciously aware of their happening. As one group member succinctly described the process, 'You get up and go without really thinking, you know exactly where you have to go, and you get there but you really don't think about getting there while you're on your way.' The phrasing of this statement in almost poetic fashion points to a kind of automatic unfolding of movement with which the group member has little or no conscious contact. She has no recollection of the great number of footsteps, turns, stops, and starts that in sum compose the walks from home to school. She finds herself at her destination without having paid the least bit of attention to the movement as it happened at the time.

Traditionally, behaviorist and cognitive theorists have dealt with habitual movement in two contrasting ways. The latter students argue that habitual behaviors are not really habitual; that if the person could actually see the inner processes directing movement, he would discover that he is consciously evaluating the situation at hand and making constant use of his cognitive map:

> Admittedly, much of spatial behavior is repetitious and habitual – in travelling, you get the feeling that 'you could do the trip blindfolded' or 'do it with your eyes shut.' But even this apparent 'stimulus-response' sequence is not so simple: you must be *ready* for the cue that tells you to 'stop now' or *evaluate* the rush hour traffic that tells you to 'take the other way home tonight.' Even in these situations you are *thinking ahead* (in both a literal and metaphorical sense) and using your cognitive map (Downs and Stea, 1973, 10).

In contrast, strict behaviorists reject any cognitive process intervening between environment and behavior.[8] They have consistently emphasized the automatic nature of everyday movement, which they define in terms of *reinforcement* – i.e. any event the occurence of which increases the probability that a stimulus will on subsequent occasions evoke a response (Hilgard *et al.*, 1971, 188-207). Applied to spatial behavior, this principle argues that a successful traversal of space over a particular route strengthens the chances that this route will be used the next time the organism must traverse that space. Each time the movement is repeated the responses evoking that particular route are *reinforced* and in time the pattern becomes habitual and thus involuntary.

In proceeding phenomenologically, the researcher must place in parentheses these two contrasting interpretations and ask what habitual movement is *in its own fashion*. Through this bracketing procedure, one sees the sensitive role that body plays in much of everyday movement.

The Notion of Body-Subject

Additional observations from the environmental experience groups suggest that the habitual nature of movement arises from the body, which houses its own special kind of purposeful sensibility. One group member described a drive to the dentist. At one intersection on the route, he suddenly found himself turning left, rather than continuing straight as he should have done. He explained that generally he *does* turn left because he has friends up the street whom he visits often. In describing the unintentional turn, he explains that something in him acted before he could cognitively act and that this

> something is a directed action *in the arms:* my arms were turning the wheel . . . they were doing it all by themselves, completely in charge of where I was going. The car was halfway through the turn before I came to my senses and realized my mistake.'

Another person pointed to this same kind of directed bodily movement when he described the act of turning on a string lightswitch: 'my hand reaches for the string, pulls, and the light is on. The hand knows exactly where to go.'

The role of the body in movement is indicated by other observations. One group member made reference to an intelligent force in her legs which gets her about: 'you let your legs do it and don't pay any attention to where you're going.' Another group member noticed that in sitting at his desk, his hands always automatically reached for a required envelope, scissors, or other object without his having to direct them consciously. A third group member described this same bodily process in his ability to place letters quickly in their proper mailboxes when he worked at a post office. Taken together, these reports indicate that underlying and guiding many everyday movements is an intentional bodily force which manifests automatically yet sensitively: an arm reaches for string or envelopes, hands turn the steeringwheel or place letters in their proper mailbox, feet carry a person automatically to his destination. Borrowing the term from the French phenomenologist Merleau-Ponty (1962), I call this bodily intentionality *body-subject.* Body-subject is the *inherent capacity of the body to direct behaviors of the person intelligently, and thus function as a special kind of subject which expresses itself in a preconscious way usually described by such words as 'automatic,' 'habitual,' 'involuntary,' and 'mechanical.'*[9]

The possibility that the body is an intelligent subject manifesting in its own special fashion is foreign to both cognitive and behaviorist theories of behavior. Both perspectives view the body as *passive* – as an inert thing which responds to either an order from cognitive consciousness or stimuli from the external environment. Studies in spatial cognition have directed their attention to the cognitive map, as it is a record of the person's cognitive knowledge of space. This research has focused little attention on the actual bodily movements which constitute spatial behavior. Instead, it has emphasized the cognitive process which is assumed to coordinate relations between environment and behavior.

On the other hand, behaviorists have emphasized the significance of body in their discussions of spatial behavior, but in this emphasis they have viewed it as a collection of reactions to external stimuli. If the behaviorist were asked, for example, to describe driving behavior from home to work, he would argue that it involves a succession of reactions to the shifting sights, sounds, and pressures impinging on driver's external sense organs, plus internal stimuli coming from viscera and skeletal muscles (Tolman, 1973). These various stimuli call out particular feet and arm movements of the driver which are reinforced each time a particular driving response successfully gets the driver safely to his destination. Eventually, this series of stimulus-responses is integrated into a smooth, step-wise progression which easily and automatically gets the person from home to work each day.

Reports from the environmental experience groups suggest that both cognitive and behaviorist theories are incomplete. The cognitive description is lacking because it ignores the fact that many movements proceed independently of any cognitive evaluation process; that the cognitive stratum of experience comes into play only when body-subject makes a wrong movement, as when, for example, the dialer *becomes aware* that he is dialing the wrong number or when the driver *realizes* that he is making the wrong turn. Otherwise, cognitive attention is directed to matters other than the behavior at hand. The driver making the wrong turn, for example, explains that his attention was on the impending visit to the dentist.

Yet the fact that cognitive attention can intervene when body-subject errs points to a first weakness of the behaviorist perspective: behavior can involve a cognitive component and is therefore *more* than a simple sequence of stimulus-response behaviors. Furthermore, the notion of body-subject calls into question the entire concept of stimulus-response, since body-subject is an intelligent, holistic process which *directs,* while for the behaviorists, the body is a collection of passive responses that can only *react.*

If one reviews the above reports, he notes no experiential indications that the movements described are a series of responses to external things in the environment. Consider the person turning on the light. The central theme in his report is the directed way in which the arm goes up: 'the hand knows exactly what to do.' The environmental context here seems almost secondary, and in fact the person explains that the arm can find the string as well in the dark as in the daylight. Similarly, the focus in the report on the wrong turn is the hands which do the turning 'all by themselves, completely in charge.' Here, too, the tone of the observation points to the hands as an intelligent agent in charge of the situation in their own special way. These observations provide no indications that the body is blindly responding to environmental stimuli as the behaviorist would assume. Rather, the reports suggest that the body acts in an intentional way which tackles the behavior needed as a whole and proceeds to carry it out in fluid, integrative fashion.

Movement, explored phenomenologically, indicates that the body is intelligently active and through this activity efficiently transforms a person's needs into behaviors. If one is to move effectively to meet the requirements of everyday living, his body must have within its ken the required habitual behaviors. Without the structure of body-subject, people would be constantly required to plan out every movement anew – to pay continuous attention to each gesture of the hand, each step of the foot, each start. Because of body-subject, people can manage routine demands automatically and so gain freedom from their everyday spaces and environments. In this way, they rise above such mundane events as getting places, finding things, performing basic gestures, and direct their creative attention to wider, more significant life-dimensions.

For behavioral geography and environmental psychology, the notion of body-subject has important implications, especially for the cognitive approach to environmental behavior. Perhaps the biggest challenge that this approach faces is to demonstrate conclusively a link between cognition and behavior. So far this demonstration is lacking. Moore (1979, 64) writes:

> the belief of all cognitively oriented researchers in the power of intervening cognitive variables has not been followed by empirical tests of sufficient scope or quality . . . [E] xcept for the few studies of market behavior and interurban migration, we have scant data on the relationship of environmental cognition and subsequent urban behavior.

Consideration of the bodily dimension of environmental behavior indicates that the cognitive perspective is incomplete and needs a thorough rephrasing. Images, subjective impressions, category systems, and cognitive maps may have a partial role in environmental behavior, but we need a better understanding of their relative importance. More than likely, there is some kind of reciprocity between body and mind, habit and wish for change, past and future. If behavioral geography is to have an accurate behavioral theory, it must give this reciprocity more attention (Seamon, 1979, 61-2).

Body and Place Choreographies

Body-subject assures that gestures and movements learned in the past will readily continue into the future. It handles the basic behaviors of everyday living. Body-subject is a stabilizing force, and through it people gain the freedom to extend their world horizons.

Reports from the environmental experience groups point to the versatility of body-subject. It also houses more complex behaviors extending over time as well as space. One such behavior is what I call *body-ballet – a set of integrated behaviors which sustain a particular task or aim,* for instance, washing dishes, plowing, housebuilding, potting, or hunting. Body-ballets are frequently an integral part of a manual skill or artistic sensibility; their sum may constitute a particular person's livelihood. 'His movements were incredible – they flowed together,' said one group member, describing a metalsmith, whom he called an 'artist.' 'Both hands were working at once . . . doing exactly what they had to do perfectly.' Another group member described his operating an ice cream truck for a summer. As he worked, he would 'get into the rhythm of getting ice cream and giving change.' He would automatically 'reach for the right container, make what the customer wanted, and take his money.' Generally, the work required little conscious attention: 'most of the time I didn't even have to think about what I was doing – it all became routine.'

These observations suggest that through training and practice, basic movements of body-subject fuse together into wider bodily patterns that provide a particular end or need. Simple arm, leg, and trunk movement become attuned to a particular line of work or action and direct themselves spontaneously to meet the requirements at hand. In using the words 'smooth,' 'flow,' and 'rhythm,' the observations indicate that body-ballet is organic and integrated rather than stepwise and fragmentary. Once having mastered the basic operations of an activity, body-subject can vary its behaviors creatively to meet quickly the particular requirement at hand.

Similar to the body-ballet, a *time-space routine* is *a set of habitual bodily behaviors which extends through a considerable portion of time.* Reports from the environmental experience groups indicate that sizable portions of a person's day may be organized around such routines. One group member described a morning routine which he followed practically every day except Sundays. He would be up at seven-thirty, make his bed, perform his morning toilet, and be out of his house by eight. He would then walk to the corner café up the street, pick up the newspaper (which *had* to be the *New York Times*), order his usual fare (one scrambled egg, toast, and coffee), and stay there until around nine when he would walk to his nearby office. A second group member described a time-space routine which her grandmother followed: 'she is always in a particular place at a particular time and usually doing a particular thing there.' Between six and nine, for example, the grandmother would be working in the kitchen; between nine and twelve, sewing in the front porch.

These reports indicate that a series of behaviors which are in themselves body-ballets (e.g. a bathroom or sewing routine) fuse into wider patterns also directed by body-subject. As the first reporter explains, he doesn't figure out his morning routine each day; rather, 'it unfolds and I follow it.' A change in routine can cause irritation: 'I really like this routine and I've noticed how I'm bothered a bit when a part of it is upset – for example, if the *Times* is sold out, or if the booths are taken and I have to sit at a counter.'

The time-space routine has a certain holistic pattern which, like movement itself, is well described by the word 'unfolding'. When a person has established a series of time-space routines in his typical daily or weekly schedule, large portions of his day can proceed with a minimum of planning and decision. The person may become attached to these routines; interference, as the above observation indicates, can generate a certain amount of stress. Time-space routines are an essential component of daily living because they appropriate activities automatically through time. They maintain a continuity in people's lives, allowing them to do automatically in the present moment what they have learned in the past. In managing the routine, repetitive aspects of daily living, time-space routines free people's cognitive attention for more significant events and needs. On the other hand, time-space routines may be difficult to break or change. In this sense they are a conservative force which may be a considerable obstacle in the face of useful progress or change.

In a supportive physical environment, time-space routines and body-ballets of the individual may fuse into a larger whole, creating a space-environment dynamic called *place-ballet.* The *place-ballet is a fusion of many time-space routines and body-ballets in terms of place.* Its result may be an environmental vitality like that found in the streets of Boston's North End or New York's Greenwich Village. It generates a strong sense of place because of its continual and regular human activity.

Familiarity arising out of routine is one ingredient of place-ballet. One group member described her job in a corner grocery store. She had got to know customers by face because several came there regularly during her hours of work. She appreciates this interaction because 'it helps time pass faster and gives me people to talk to.' Similarly, the group member describing his morning routine explained that there are other regular patrons there at the same time as him each morning. Their presence each day generates a sense of friendliness and familiarity which he feels 'wouldn't be there if they were new faces each day.' In these examples, individual routines meet together in terms of place. The regularity so generated may also produce a climate of familiarity which participants appreciate and to which they grow attached. This regularity is unintentional and only comes about through time and many repeated 'accidental' meetings. At its base is the habitual force of body-subject, which supports a continuity grounded on bodily patterns of the past.

Wider Contexts

For geography and other disciplines of environment and place, the notions of body-ballet, time-space routine, and place-ballet have value because they join people with space, place, and time. Though the above examples are limited and culture-bound, their underlying experiential patterns transcend particular social and temporal contexts and can be found in all human situations, past and present, Western and non-Western. Hockett (1973, 13-14), for example, has pictured the typical daily routine of the Menomini, an Indian tribe living along the north-western shore of Lake Michigan in the seventeenth century. The women rose at dawn to fetch water, build the fire and prepare breakfast, which was one of two regular daily meals. After breakfast the men and boys go to the hunting and fishing grounds while the women tend the crops, process food, gather edible plants, weave, and care for the children.

Time-space routines and body-ballets are the foundation of this typical daily pattern. The activities follow a sequence which is largely habitual and unpremeditated. The women's activities are an extended time-space routine incorporating many individual body-ballets – water fetching, fire building, crop tending, and weaving. Each activity requires a particular combination of gestures and movements which correctly manipulate materials at hand and produce the desired artifact or aim. The skill of weaving, for example, is a knowledge of the hands, which long ago learned a proper sequence and rhythm and can now conduct their work quickly and automatically.

One can also visualize in the above scene a series of place-ballets unfolding throughout the Menomini's day. He can imagine, for example, the women's meeting at the stream as they fetch water and partaking in conversation. This place is not only a source of water but a scene of community interaction and communication which repeats each morning because of the regularity of water fetching. The underlying structure of this place-ballet is no different from the contemporary street scene that Jane Jacobs describes on the block where she once lived in Greenwich Village in New York:

> The stretch of Hudson Street where I live is each day the scene of an intricate sidewalk ballet. I make my own first entrance into it a little after eight when I put out the garbage can, surely a prosaic occupation, but I enjoy my part, my little clang, as the droves of junior high school students walk by the center of the stage dropping candy wrappers . . . While I sweep up the wrappers I watch the other rituals

> of the morning: Mr. Halpart unhooking the laundry's handcart from its mooring to a cellar door, Joe Cornacchia's son-in-law stacking out the empty crates from the delicatessen, the barber bringing out his sidewalk folding chair, Mr. Goldstein arranging the coils of wire which proclaim the hardware store is open, the wife of the tenement's superintendent depositing her chucky three-year-old with a toy mandolin on the stoop, the vantage point from which he is learning the English his mother cannot speak. Now the primary children, heading for St. Luke's, dribble through to the west, and the children from P. S. 41, heading toward the east (Jacobs, 1961, 52-3).

The essential experiential process working on Hudson Street, in the Menomini village, in the corner café is much the same, though on the surface each place is considerably different from the others. People come together in time and space as each individual is involved in his or her own time-space routines and body-ballets. They recognize each other and often partake in conversation. Out of these daily, taken-for-granted interpersonal dynamics, these spaces of activity evolve a sense of place that each person does his small part in creating and sustaining.

These places are more than locations and space to be traversed. Each comes to house a dynamism which has arisen naturally without directed intervention. These spaces taken on the quality which Relph (1976, 55) has called *existential insidedness* – a situation in which 'a place is experienced without deliberate and selfconscious reflection yet is full with significances.' Relph goes on to say that existential insidedness is the very foundation of the experience of place, and this point is echoed in the place-ballet. Through habitual patterns meeting in time and space, an area can become a place shared by the people who come into spatio-temporal contact there. The dynamism of that place is largely in proportion to the number of people who share in its space and thereby create and share in its tempo and vitality.

Implications

If the geographer is to study the spaces, places, and environments in which a person typically lives and dwells – his *lived-space,* as phenomenologists sometimes speak of it (Bollnow, 1967) – he must recognize that this space is first of all grounded in the body. Through body-subject, the person knows where he is in relation to the familiar objects, places, and environments which in sum constitute his everyday geographical world.

Whatever the particular historical and cultural context, the bedrock of his geographical experience is the prereflective bodily stratum of his life – his bodily lived-space. As well as habitual movements in larger-scaled environments, this bodily lived-space incorporates smaller gestures such as stepping, turning, reaching, and the extended patterns of body-ballet and time-space routine. By exploring the bodily portion of lived-space, the geographer gains a picture of the stabilizing, habitual forces of a particular lifeworld. He can better understand how unselfconscious patterns within a particular place continue to make the place what it was in the past. Further, he may be better able to predict the effect of particular environmental or social changes on that stability.

A strong pattern in modern society is the fragmentation of space and time: home is separated from market, neighborhoods are split by expressways, work is separated from leisure. At the same time, social critics speak of growing alienation and the gradual breakdown of community, which Slater (1970, 5) has defined as 'the wish to live in trust and fraternal cooperation with one's fellows in a total and visible collective entity.' If community is an important component of a satisfying human existence and if community is presently eroding, the geographer can well ask what community is in experiential terms and how it might be buttressed. Place-ballet may have considerable bearing on this question because it regularly brings people face to face who otherwise would probably not know each other. In this sense, place-ballet generates some of the interpersonal cooperation and trust of which Slater speaks.

Place-ballet has particular bearing on the nature of neighborhood, which the economist Barbara Ward once defined as a place where

> children can grow up without being run over, where friends can meet, where that deeply reglected resource, two human legs, can be recovered and used, and where the sociability and the exchanges of human existence can take place in a civilized form (cited in Horsley, 1978, 1-2).

In one sense, this definition of neighborhood is founded in place-ballet. Ward suggests a community in place grounded in a bodily scale and interpersonal continuity. In part, such a place must be founded in familiarity: people knowing each other well enough so they can comfortably interact. A portion of this familiarity may well be the result of unselfconscious regularity. A habitual base, however, does not mean a precisely predictable neighborhood dynamic composed of robot-like humans continuously repeating the same sets of behaviors. Rather, the precognitive regularity of place-ballet provides a foundation from which can arise surprise, novelty, and unexpectedness: the spontaneity of childplay, neighbors 'bumping into' one another, a community group quickly organizing to oppose a proposed street widening. Stability and continuity of place, therefore, are at least partially responsible for the civilized network of interactions that Ward points to in her definition of neighborhood. At the same time, this order in terms of place establishes a pattern of regularities around which a progression of shifting events and episodes can occur. Place, in other words, requires both regularity and variety, order and change. Place-ballet is one means by which a place comes to hold these qualities.

Neighborhood is only one example of place-ballet. Any situation where at least some users come together regularly – for example, lounge, café, office building, marketplace – may provide a base for place-ballet. At the same time, however, many place-ballets appear to be eroding and disappearing. The trend, as Relph (1976, 117) reminds us, 'is toward an environment of few significant places – toward a placeless geography, a flatscape, a meaningless pattern of buildings.'

In this sense, the notion of place-ballet has important theoretical and practical implications. First, it joins people, time, and place in an organic whole and portrays place as a distinct and authentic entity in its own right. In the past, many approaches to the person-environment relationship have been piecemeal and mechanistic: place is only the sum of the behaviors of its individual human parts. In contrast, place-ballet depicts a whole greater than its elements: place is a dynamic entity with an identity as distinct as the individual people and environmental elements comprising that place. Place-ballet, in other words, is an environmental synergy in which human and material parts unintentionally foster a larger whole with its own special rhythm and character. An outdoor marketplace, for example, may be grounded in economic transactions, but is considerably more than just those transactions (Nordin, 1976). The market takes on an atmosphere of vitality, camaraderie, excitement – even gaiety. 'Meeting friends and acquaintances' becomes as important as buying (Seamon and Nordin, 1980).

For environmental planning, therefore, place-ballet provides a notion around which to construct policy and design. What places have what kind of place-ballets? Would a place be a better human environment if it had place-ballets? How could place-ballets be started in that place? For residents of places themselves, the notion of place-ballet is especially valuable. People recognize unconsciously the significance of place-ballet but generally because of the natural attitude have no refined means for articulating the entity in clear terms. Place-ballet makes one implicit dimension of the lifeworld explicit. It provides an articulated concept which might be of value in creating, regenerating, and protecting places.[10]

Notes

1. On phenomenology, its history and method, see Natanson, 1962; Wilde, 1963; Spiegelberg, 1965. Examples of empirical phenomenology can be found in Giorgi *et al.*, 1971, 1975.
2. In practice this definition discounts reflexive movements such as blinking, breathing, etc.
3. On behaviorism, see Taylor, 1967; Merleau-Ponty, 1962, 1963; Koch, 1964. There is not one behaviorist theory but many. Some behaviorists recognize cognition as a significant intervening process in the stimulus-response sequence. See, for example, Tolman, 1973.
4. See Hull, 1952; Blaut and Stea, 1973; Getis and Boots, 1971.
5. E.g. Lynch, 1961; Stea and Downs (eds.), 1973; Moore and Golledge (eds.), 1976; Leff, 1977.
6. A full account of this group process is provided in Seamon, 1979.
7. Complete statements of these and following accounts can be found in Seamon, 1979.
8. Again, it is important to realize that some behaviorists interpret cognition as an important intervening process between stimulus and response. See note 3.
9. Merleau-Ponty, 1962. For further discussion of Merleau-Ponty's significance here, see Seamon, 1979.
10. This possibility is considered further in Seamon, 1979, Chapter 19.

References

Bollnow, Otto. 1967. 'Lived-space.' In N. Lawrence and D. O'Connor (eds.), *Readings in Existential Phenomenology,* pp. 178-86. Englewood Cliffs, NJ: Prentice-Hall.

Downs, Roger M., and Stea, David (eds.). 1973. *Image and Environment: Cognitive Mapping and Spatial Behavior.* Chicago: Aldine.

Getis, Arthur, and Boots, Barry M. 1971. 'Spatial Behavior: Rats and Men,' *The Professional Geographer, 23,* 11-14.

Giorgi, Amedeo. 1970. *Psychology as a Human Science: A Phenomenologically Based Approach.* New York: Harper & Row.

Giorgi, Amedeo, Fischer, W., and Von Eckartsberg, R. (eds.). 1971. *Duquesne Studies in Phenomenological Psychology,* vol. 1. Pittsburgh: Duquesne University Press.

Giorgi, Amedeo, Fischer, C., and Murray, E. (eds.), 1975. *Duquesne Studies in Phenomenological Psychology,* vol. 2. Pittsburgh: Duquesne University Press.

Hilgard, E.R., Atkinson, R., and Atkinson, L. 1974. *Introduction to Psychology.* New York: Harcourt Brace Jovanovich.

Hockett, C.F. 1973. *Man's Place in Nature.* New York: McGraw-Hill.

Horsley, C.B. 1978. 'Making of a City Neighborhood,' *New York Times,* 24 December, Secs. 8-9, pp. 1-2.

Hull, Clark. 1952. *A Behavior System.* New Haven: Yale University Press.

Jacobs, Jane. 1961. *The Death and Life of Great American Cities.* New York: Vintage.

Koch, Sigmund. 1964. 'Psychology and Emerging Conceptions of Knowledge as Unitary.' In T.W. Wann (ed.), *Behaviorism and Phenomenology,* pp. 1-45. Chicago: University of Chicago Press.

Leff, Herbert. 1977. *Experience, Environment, and Human Potentials.* New York: Oxford University Press.

Lynch, Kevin. 1960. *The Image of the City.* Cambridge, Mass.: MIT Press.

Merleau-Ponty, Maurice. 1962. *Phenomenology of Perception,* trans. Colin Smith. New York: Humanities Press.

—, 1963. *The Structure of Behavior,* trans. A.L. Fisk. Boston: Beacon Press.

Moore, Gary T. 1979. 'Knowing about Environmental Knowing: the Current State of Theory and Research on Environmental Cognition,' *Environment and Behavior, 11,* 33-70.

Moore, Gary T., and Golledge, R.G. (eds.). 1976. *Environmental Knowing: Theories, Research, and Methods.* Stroudsburg, Pa.: Dowden, Hutchison, and Ross.

Natanson, Maurice. 1962. 'Phenomenology: a Viewing.' In *Literature, Philosophy, and the Social Sciences,* pp. 3-25. The Hague: Martinus Nijhoff.

Nordin, Christina. 1976. 'Varbergs Torg i Tiden.' In *Varbergs Museum-Årsbok 1976,* pp. 111-61.

Seamon, David. 1979. *A Geography of the Lifeworld: Movement, Rest and Encounter.* London: Croom Helm; New York: St Martin's Press.

Seamon, David and Nordin, Christina, 1980. 'Market Place as Place Ballet: The Example of Varberg, Sweden,' *Landscape*, 24 (forthcoming).

Slater, Philip. 1970. *The Pursuit of Loneliness: American Culture at the Breaking Point.* Boston: Beacon Press.

Spiegelberg, Herbert. 1971. *The Phenomenological Movement: an Historical Introduction,* vols. 1 and 2. The Hague: Martinus Nijhoff.

Stea, David, and Blaut, James M. 1973. 'Notes Toward a Developmental Theory of Spatial Learning.' In Downs and Stea (eds.) (1973), pp. 51-62.

Taylor, Charles. 1967. 'Psychological Behaviorism.' In *The Encyclopedia of Philosophy,* vol. 1, pp. 516-20. New York: Macmillan and the Free Press.

Tolman, C. 1973. 'Cognitive Maps in Rats and Men.' In Downs and Stea (eds.) (1973), pp. 27-50; originally in *Psychological Review, 55* (1948), 189-208.

Wilde, John. 1963. *Existence and the World of Freedom.* Englewood Cliffs, NJ: Prentice-Hall.

Zeitlin, Irving M. 1973. *Rethinking Sociology.* New York: Appleton-Century-Crofts.

8 HOME, REACH, AND THE SENSE OF PLACE*

Anne Buttimer

Country Road, take me home
to the place
I belong . . .
John Denver.

Emotionally laden eulogy on the meaning of place rings through much modern poetry and song. Nostalgia for some real or imagined state of harmony and centeredness once experienced in rural settings haunts the victim of mobile and fragmented urban milieux. Like many a fortune-seeker amidst the lights of Broadway who longed for the simple cabin near the rippling stream back home I suppose one could say, 'You never know what you've got 'til it's gone.' Patriotic songs about native soil and forest that built the spirit of nationhood in European countries were often written in the cities of North America and Australia. And today, as the uniqueness of places becomes more and more threatened by the homogenizing veneer of commercialism and standardized-component architecture, many people long for their *hembygd* and *smultronställe*.

It is fascinating to notice when and where during recent history this notion of place has emerged as a strong motif in literature, politics, and popular song. The record synchronizes fairly well with periods of relatively abrupt change either within the social or physical environment or in the world of ideas. Late-eighteenth-century and early-nineteenth-century Romantic literature on place, for instance, corresponds roughly with the reaction against a Newtonian world view. Scandalous, it seemed, to impose a 'scientific' grid on Nature – to reduce beauty, melody, and fragrance to the sterile metric of mathematics or physics.

When industrialism and transport systems began to rupture the old harmonies of peasant landscapes, again protest was voiced in the language of place. Urbanization brought its own wave of rebellion against abrupt change: the old mosaic of artisan districts, open markets, and bourgeois villas became distorted and dismantled as within the city itself former cultural and economic equilibria gave way to the new. 'Housekeeping'

* A slightly different version of this paper originally appeared in *Regional identitet och förändring i den regionala samverkans samhälle* (Acta Universitatis Upsaliensis, Uppsala, Sweden, 1978), pp. 13-39.

functions of civic life – residential, ceremonial, governmental, and religious – yielded to the growing importance of 'empire-building' functions, e.g. commerce, finance, and industry. In the bustling enthusiasm of early industrialism it was far more important to expand horizons of access to markets and clientele than to try seriously to make the city a home.

Transatlantic migrations in the late nineteenth and early twentieth centuries mark another powerful source of lament and insight into the meaning of place. European migrants often sang praises of their home places. To those for whom the westbound Oceanic voyage was final, the longing for home resulted in a virtual torrent of feelings about places and their identity. Many groups had indeed tried to incarnate their images of 'home' not only in social and political life but also in their choices of work, living, and recreation patterns. Even today 'Little Italys,' 'Bohemias' and 'Blarney Stone Irish' neighborhoods of American cities still display façades so 'authentic' that the European visitor is reminded of his grandparents' era.

Changing Approaches to Place

Whatever its sources of explanation, this literature on the sense of place reveals several consistently recurring themes. It appears that people's sense of both personal and cultural identity is intimately bound up with place identity. Loss of home or 'losing one's place' may often trigger an identity crisis. In his *Poetics of Space,* the twentieth-century philosopher Bachelard (1958) claimed that the relationship between place and personality is so intimate that to understand oneself a *topoanalysis* – the exploration of self-identity through place – might yield more fruitful insight than *psychoanalysis.* There are many dimensions to meanings ascribed to place: symbolic, emotional, cultural, political, and biological. People have not only intellectual, imaginary, and symbolic conceptions of place, but also personal and social associations with place-based networks of interaction and affiliation. As with other members of the biosphere, too, humans display marked patterns of territoriality. When the fundamental values associated with any of these levels of experience are threatened, then protest about the meaning of place may erupt. Whether all these values are *consciously* articulated in legal or behavioral terms does not seem to be the crucial point. In fact, they are often not brought to consciousness until they are threatened: normally, they are part of the fabric of everyday life and its taken-for-granted routines.

This does not suggest, however, that ideas about place are not significant. In fact, the fundamental *Zeitgeist* or world view at any point in history which may be explicit in scientific and philosophical ideas about space, time, and nature is present implicitly also in conceptions of place. It is especially on this level that the twentieth century brought unprecedented reaction and insight. The 'post-Newtonian shock' was over: science and rationality had triumphed over other all-competing alternatives. That countries and places should be planned within a wider socio-spatial horizon was taken for granted. World depression and war justified managerial convictions that people and their home places should no longer be trusted to carry on in the traditional way. An old stoic idea that rational order should be imposed on nature and society became an apparently workable dream because of developments in science and technology.

As transport and communication systems shrank the distances between places and the increasing mobility of people, jobs, and armies continued to level out differences between places; it was naturally in the urban context that the question of local identity became politically articulate. Prototypical examples of the place-based 'culture shock' are the ethnic ghettoes of the early twentieth century in North America. Migrants from rural backgrounds in Europe poured into the industrial cities of New England, Chicago, and the West Coast, all eager to maintain some sense of cultural and kin identity while joining the queue for the unskilled job market. These pockets of first-generation immigrant settlement quickly assumed a distinctive 'ethnic' character. Usually circumscribed by either physical landmarks, zoning laws, or prejudice, they displayed an internal cohesion reminiscent in many ways of working-class districts in European cities. Social scientists became fascinated by the social order of the slum: they noted the interweaving of social and spatial arrangements and ways in which external landscape forms mirrored the world views and behavior patterns of their residents. With confidence indeed did the Chicago Municipal Authorities accept the recommendations for administrative boundaries offered by these community sociologists: boundaries which still persist despite massive demographic and cultural change. The lack of 'fit' between administrative regionalization plans and the social character of local areas remains an enduring problem regarding the identity of place.

Prior to World War II most of what was written about place was construed in terms of a Newtonian world: container space for people and activities, areally circumscribed domains of political authority and administration. The notion of place was still a very respectable one: credible to those who lived in them as well as to those who sought to plan them. Place had served as a model for the resettlement of minorities after World War I: the boundaries of language and nationalism being understood as basic and unquestionable (Dominian, 1920). Place was also a plausible concept for two key planning movements of the 1920s: the Garden City movement in Britain (Howard, 1897) and the famous *Regional Plan for New York and its Environs* in the USA (Haig, 1927, 1929).

It was only after World War II that the full impact of an emerging Einsteinian world view began to express itself in applied science and its reverberations on character of places. National and multi-national economic planning became a moral imperative: political boundaries were dismantled and recombined in various ways; comparative advantages in production and delivery were sought; optimal spatial and structural programmes were designed to maximize efficiency and interaction. The sky was the limit, it seemed, on the advantages of agglomeration and scale economies. Most of these movements triggered significant popular reactions. Perhaps most striking is the way in which national stereotypes persist, for example, among the member countries of the European Economic Community, despite the enormous changes in their actual patterns of interaction and exchange. But it was in the urban realm especially that the old refrain of place identity assumed the most dramatic forms.

Urban renewal programmes, particularly in the United Kingdom and the United States, nearly always found their first targets in 'slum' areas near the old city center – victims of the Central Business District's success as dominant pole within the empire-building city which had forgotten its housekeeping responsibilities. Bulldozers respected few of those invisible boundaries or sacred symbols of social space. (Few there were of course: in Glasgow today, several years after the first bulldozer arrived, pubs and churches stand like *mesas* atop a landscape razed to flatness – isolated from their clientele whose houses once lay around and above them.)

When social scientists got involved as consultants either to redevelopment plans or to their evaluation, they usually brought along the older models of place identity which had been tried and tested in the twenties and thirties. Myths of 'community' and 'territoriality' had enjoyed a long popularity. It appeared ideologically desirable for many political authorities to sponsor such people-oriented research. Most of this literature took the form of postmortem autopsy – 'grieving for a lost home' – too late or perhaps too 'safe' to influence the onward march of urban and regional development planning.

What emerged for many of the authors and audience of such studies, however, was the strong conviction that there is a fundamental contrast between the *insider's* ways of experiencing place and the *outsider's* conventional ways of describing them. Lived space and lived time were poorly and only partially represented via disciplinary models of representational space and time. For many serious researchers the road ahead lay in exploring the lived worlds of people in place.

Critical philosophers have accused those who took such an existentialist approach of having fallen prey to manipulation by ideological and managerial interests (Adorno, 1973). Beneath the jargon of authenticity, it was noted, lurked a tacit condoning of poverty, injustice, and alienation. Romantic descriptions of neighborhood life may indeed unwittingly have served to perpetuate myths like the 'culture of poverty' or to trivialize the insider's perspective by refusing to articulate it in a language which would reveal the extent of structural inequity built into contemporary economic and political life. Herein perhaps lies one basic clue to the impasse: the language used to describe the residents' perspectives on place is still, by and large, the language of a Newtonian world – people, activities, and things contained within place – whereas the language used to plan the economic and technological horizons of place has been profoundly influenced by Einsteinian conceptions of topological space, time, and process. To speak of 'insiders' and 'outsiders,' places versus spatio-temporal organization, and other dualisms of this sort may serve reasonably well to describe the historical record. But to do justice to the fundamental life interests which could be evoked by the question of place identity today, one needs to probe to a deeper level of meaning, there hopefully to find some common denominators for a dialogue between those who wish to live in places and those who wish to plan for them.

Home and Horizons of Reach

I suggest we think about places in the context of two reciprocal movements which can be observed among most living forms: like breathing in and out, most life forms need a *home* and *horizons of reach* outward from that home. The lived reciprocity of rest and movement, territory and range, security and adventure, housekeeping and husbandry, community building and social organization – these experiences may be universal among the inhabitants of Planet Earth. Whether one thinks on the level of ideas themselves, or of social networks, or of 'home grounds,' there may be a manner in which one can measure and study the reciprocity of home and reach in all of them. For any individual the home and reach of one's thought and imagination may be quite distinct from the home and reach of one's social affiliations, which may again be distinct from the actual physical location of physical home and reach (Rose, 1977). These distinctions are not just abstractions: if they are actually mapped within the lived-space time horizons of any individual or group they could provide some clues into what constitutes place identity. If all three are synchronized or harmonized then one could speak of *centeredness* and hypothesize that one's sense of place is a function of how well it provides a center for one's life interests. Taken in a more general sense, the question becomes how many of a local area's life interests may be centered within it and how many of them have their 'home' elsewhere?

This process of centering is not at all identical with the notion of *centralization* – the rationally planned nodal concentration of power and social energy. Rather, centering suggests an ongoing life process – the breathing in and bringing home which is a reciprocal of the breathing out and reaching toward horizon. The essential difference between centralization and centering may in fact symbolize the language between insiders' and outsiders' views on place. For the words used to describe places *looked at from the outside* are nouns – artifacts like housing, land use, activity flows, political boundaries. From a national perspective, one may speak of centralizing or decentralizing them. In contrast, centering is an essentially creative process authored by people themselves. The meanings of place to those who live in them have more to do with everyday living and doing rather than thinking.

To discuss place, we have to freeze the dynamic process at an imaginary moment in order to take the still picture. The observer who explores place speaks of housing, whereas the resident of that place lives the process of dwelling. The observer measures and maps activity systems and social networks and infers something of the native's world within reach, whereas in the resident's experience reaching may be so fundamental a movement of everyday existence that it is not usually reflected upon. One of the first steps to be taken in trying to straddle the divide between insider and outsider worlds, then, is to stretch our conventional 'noun' or 'picture' language so as to accommodate the 'verbs' and 'process' languages of lived experience.

The 'outsider's trap,' to exaggerate a bit, is that one looks at places, as it were, from an abstract sky. He or she tries to read the texts of landscapes and overt behavior in the picture languages of maps and models and is therefore inevitably drawn toward finding in places what he or she *intends* to find in them. The 'insider's trap,' on the other hand, is that one lives in places and may be so immersed in the particulars of everyday life and action that he or she may see no point in questioning the taken-for-granted or in seeing home in its wider spatial or social context. For both insider and outsider, perhaps the greatest challenge is pedagogical: a calling to conscious awareness those taken-for-granted ideas and practices within one's personal world and then to reach beyond them toward a more reasonable and mutually respectful dialogue.

The geographer, of all people, should find such a pedagogical challenge attractive. Whereas the bulk of our disciplinary models have fostered an observer's stance on places, we all sometime somewhere are insiders in some place. Would it not be a worthwhile goal, then, to examine our own experiences in various places and use this as a testing ground for our generalizations and also our efforts to achieve better communication across this divide? Before one leaps into issues of planning policy it would be vital, it seems to me, to understand the fundamental life processes which are at stake and are vulnerable to changes in the physical and political identity of place. The question of decentralization, for instance, might be redefined as a problem that calls for an effort to revitalize the energies of local areas from the grass roots at the same time as one studies the feasibility of blueprints for redistribution from the top down.

Experiences in Place

The rest of this essay offers some insights and questions arising from exploratory efforts to understand the meaning of place in my own experience as well as in that of other groups. This is not intended as a model to be tested or a solution to problems but rather the story of a journey across the divide which most of us confront (or ignore) when we speak of people and places. In a way it is only the kind of 'homework' which any scholar who believes in objectivity should be willing to face: to unearth the 'subjective' element and recognize its influence.

I'm sure that many of the attitudes I bring to my geography, for instance, and certainly my cynicism about top-down bureaucratic planning, derive from my childhood experiences of life in Ireland (Figure 8.1). It is difficult for me to find words to describe what experience of living in Ireland still means to me. It is a total experience of milieu which is evoked: I recall the feel of the grass on bare feet, the smells and sounds of various seasons, the places and times I meet friends on walks, the daily ebb and flow of milking time, meals, reading and thinking, sleeping and waking. Most of this experience is not consciously processed through my head – that is why words are so hard to find – for this place allows head and heart, body and spirit, imagination and will to become harmonized and creative. So far, Mansholt-type rationalization has not disrupted the fabric of our local landscape or the predictability of sonic and other intrusions throughout the day; somehow to live there allows one a sense of being in tune with the rhythmicity of nature's light and dark, warmth and cold, sowing and harvesting (Figure 8.2).

Now as a geographer I could probably tell you what soil types there are, what the crop rotation system used to be, the time-cost-distance to markets and so on, but even when I had done the round of 'geographical'

Figure 8.1: Glenville, Co. Cork, Ireland

Figure 8.2: Glenville, Co. Cork, Ireland

interpretations I still would feel it was only an opaque picture – insufficient to tell you whether the administrative/economic/political boundaries of place should be revised or what realms of human life interests should be left to our discretion or arrogated by regional or national authorities. Perhaps here I'm encased in the 'insider's trap'? Experiences in several other environments, however, have convinced me that the solutions to many regional problems which have been discovered through trial and error over generations may still be more rational than those conceived by the urban or foreign civil servant. Experience also suggests that any solution which people do not consider to be 'their idea' will be resented, avoided, or rendered ludicrous over time. Attuned as I was to a relatively stable rural world with its taken-for-granted 'place' for everything, the urban landscapes of North America seemed at first to lack identity (Figure 8.3). Suburbia especially seemed such a 'placeless' sprawl of concrete and plastic nowhereland gridiron (Figure 8.4). Even in university districts, one searched in vain for that kind of sidewalk café where one could linger for hours: rather one found the short-order eating places which prided themselves on quick turnover and throughput (Figure 8.5). It did not take long to discover how false it could be to judge the book by the cover: many of these nondescript hamburger stations were meaningful nodes in teenager and truckdriver social space.

To many people, for whom *reaching* appears to be more important than *home making,* places may be simply points on a topological surface of access. Eventually, the architectural monotony or discordance within any one physical place could be the result of coincidence rather than design: each component is part of a spatial linkage and the center of each lies at a national or multinational headquarters. How things will eventually combine within any particular geographical milieu, therefore, is quite incidental both to those who design them and those who will use them (Figure 8.6). The skyscrapers, airports, freeways, and other stereotypical components of modern landscapes – are they not the sacred symbols of a civilization which has deified reach and derided home (Figure 8.7)? Or the gaping wounds of mining and industrial landscapes, are they not the refuse heaps of a civilisation intoxicated with Promethean hubris (Figure 8.8)?

Styles of Life and Place

It is difficult for a native of rural Ireland to take a culturally relativist attitude toward the placeless landscapes of reach (Relph, 1976; Clay,

Figure 8.3: Downtown Toronto

Source: Courtesy of Dr Hans Aldskogius, Uppsala.

Figure 8.4: Suburbia, Chicago Region

Source: Courtesy of Dr Torsten Hagerstränd, Lund.

Figure 8.5: McDonald's Main Street, Worcester (or Anywhere)

Source: Courtesy of Rudi Hartmann, Worcester, USA.

Figure 8.6: White City (Discothèque), Route 9, Mass., USA

Source: Courtesy of Rudi Hartmann, Worcester, USA.

Figure 8.7: Freeway Access to Downtown Minneapolis

Source: Courtesy of Dr Hans Aldskogius

Figure 8.8: Open-pit Mine, Minnesota

Source: Courtesy of Dr Hans Aldskogius

1974). This is not because of aesthetics alone or sensory overload; rather it derives from a nausea about values which make machines, commodities, movement and salesmanship more important than human encounter or letting nature have some breathing space. It is even hard to be wholeheartedly in favor of restoration movements in parts of cities if the aim is simply to reconstitute a museum for the benefit of the tourist industry or to salve the consciences of societies which have trampled on history: it is the style of life associated with place which is still far more important for me than its external forms (Figure 8.9).

Figure 8.9: 'Old' Montreal

The 'placing' of objects and activities within interior space was also a surprise to someone socialized within an Irish rural home. Because of the American emphasis on consumerism, it has become not only stylish but also profitable to design interior space in ways that will appeal to prospective buyers. Proxemics and territoriality – inside and outside the home – tend to become academically interesting research topics at periods of collective claustrophobia or political commitment to the expansion of *Lebensraum* – like the late 1960s on both sides of the Atlantic. It was in Germany that the special field of *Bürolandschaft* first appeared – promise of scientific bases for designing furniture, acoustics and aesthetics in ways to facilitate optimal work efficiency. In most of our 'developed' urban societies what appeals after a certain threshold of media massage has been reached is quite frequently what pays – either in terms of commercial profit or ideological consensus.

Perhaps the applied ideological or commercial fringe of territoriality study has overshadowed the deeper issues. Norberg-Shulz (1971), in his lucid textbook on architecture, outlines what is involved in making architectural space a reflection of existential space. I am, of course, doubtful that all significant aspects of existence in place can be readily translatable into architectural terms, but this does not really seem to be the important point. The crucial philosophical and pragmatic problem lies in what potential role is allowed for residents to have any creative say in designing places.

Looking back, I suspect that my decision to accept an invitation to join a team of experts on a critique of planning standards in the United Kingdom probably stemmed from an emotional resentment against the ways in which benign technocrats and planners seemed to be actually killing places through their very efforts to 'renew' them (Figure 8.10 8.11). It was this resentment plus the hope that somehow there could be some place for *reason* in the highly *rational* world of liberal planning much more than intellectual curiosity about applied geography that motivated me to investigate the lived experience worlds of working-class families. Equipped with the analytical and theoretical languages of social science on the one hand, and deeply aware of how traumatic a change of home can be, I felt that I could perhaps play a mediator role between insiders and outsiders in this context (Buttimer, 1972, and reprinted in this volume).

The details of this study are perhaps less interesting than the insight which later reflection yielded regarding its strengths and limitations. There were several ways in which the actual models used prevented me from letting the residents articulate their own versions of the situation or from communicating directly to planning authorities themselves. Also, from an analytical viewpoint there were several dimensions missing: analyses of perceptions, behavior, territoriality, and attitudes in cross-sectional fashion at one point in time did not allow me to take sufficient account of the temporal dimension. It was clear that those habits and preferences which had the deepest roots in people's memories were also those which were likely to survive and many of those were not really plannable. It was also a anthropocentric study: even environment was defined in 'human' terms (architectural form, access horizons, functional boundaries, etc.) and insufficient attention was given to the actual physical environment.

Figure 8.10: Before 'Renewal' in Glasgow Gorbals (*c.* 1965), Scotland

Figure 8.11: After 'Renewal' in Glasgow Gorbals (Hutchesontown, *c.* 1970)

But most significantly in retrospect, the study was not an objective interpretation. Even though residents spoke about 'home ground' and 'sense of place,' I heard those words through the filters of my own experience. 'Home' for me should ideally have those qualities of my own home – quietness, fragrance, spaciousness, rhythmic flow of light and dark, winter and spring. These Glasgow housewives would probably have felt uncomfortable in such a milieu. For them the noise and bustle of street life, the regular whistle of factory and train, the occasional gang fight and football match – these would probably have been more important than the presence of cows or birds (Figures 8.12 and 8.13).

Conscious of the ways in which disciplinary models and one's own socialization can influence one's approach to the study of place, I have tried over the past few years to develop a method of investigating my own experience of home and reach in the two contrasting milieux where most of my time has been spent: the Glenville home and an apartment building on Main Street, Worcester, Massachusetts. The aim was to stand as objectively as possible inside my own experience and through sharing an inquiry with those who lived in the same apartment building to arrive at some understanding of the insider's world there.

This study has not altogether been successful. First, from an analytical perspective, it was virtually impossible to assemble data on the various relevant categories of the environment which were in any way comparable or easily relatable to the human experience of milieu. Second, although a good number of my fellow residents were elderly and retired, they scarcely ever thought about place at all. For them the processes of home and reach had primarily a social meaning: telephone, taxi, and mail kept them happily involved in their non-place-based networks. There was only a small range of activities which could be affected by the local physical environment: waiting for the bus, going to shop, church, library. They had become much better adapted to placelessness and individualism than I. This is probably prototypical of the apartment complex which sits on a main throughfare within the contemporary city.

Yet when we probed to the roots of our recurrent complaints and feelings of unease, we discovered some common denominators of concern about 'gatekeepers' to our immediate milieu. Chief among these was the landlord, in whose hands rested all discretion over details like electricity, gas, water, plumbing, and protection from vandalism. He was inaccessible most of the time even by telephone. Other gatekeepers on our immediate environment, like police and fire departments, were similarly unpredictable in the time-distance of access. Most of what actually went on within the zone of immediate reach was a function of remote

Figure 8.12: 'Home' in Drumchapel, Glasgow

Figure 8.13: 'Home' in Rural Cork

Figure 8.14: 'Housing' at 1039 Main Street, Worcester, Mass., USA

Source: Courtesy of Rudi Hartmann, Worcester, USA.

interests for which our street was but a thoroughfare. But most seriously of all, we were denied opportunity to express any responsibility for the upkeep of our place or for creating there any sense of community through mutual sharing. We were forced to adopt an attitude of surviving as individuals and thinking only of our own horizons of social reach and to blunting our sensitivities to other dimensions of reach within our own experience or being sensitive to those of others. It is hard to think about the city as a whole or engage in civic life if one's everyday *genre de vie* becomes so preoccupied with individual survival.

Perhaps this is the most serious long-term consequence of the demise of place: if one cannot practice a giving and receiving in one's normal everyday routines, is society as a whole not denied the precious input which only individuals *in situ* could contribute? Jane Jacobs' account of Hudson Street (1961) would suggest that indeed it is. An exploratory study of crime patterns along our stretch of Main Street suggested a spatia correspondence between incidence of crime and the presence of resident owners. This possibility echoes Oscar Newman's *Defensible Space* (1973), which argues that place-based round-the-clock activities may be a better deterrent to crime than police surveillance.

An Education in Place

To consider the demise of place and its consequences for personal and community life as the result of fascist decisions to centralize everything may make attractive rhetoric but not the most helpful explanation in the long run. Rather it may be pedagogically more provocative and practically more feasible to design exercises which could help insiders within their everyday milieux to become aware of the long-term implications of an individualistic and fragmented life-style both for the quality of their own lives and the general character of their residential and work environments. Such consciousness-raising efforts would be of little avail, however, if there were not a simultaneous attempt on the part of managerial interests to become educated themselves: to sort out subjective bias and role-constrained decision-making from whatever *reason* would require for the good of the whole. A geographer, sensitized to insider's and outsider's experiences of place, and aware of their reciprocity of home and reach within his or her own life experience, could surely offer some help on such a pedagogical venture (Seamon, 1979).

Returning to my Irish home place now I feel strongly about the kind of education required. There is needed an ongoing attitude of self-awareness which would help people assess the meaning of their vastly expanded horizons of reach. The old rules of thumb on 'housekeeping' may be appropriate in some realms, but they are hopelessly inadequate in others. What seems technologically desirable in some realms can be socially and ecologically disastrous in other areas. Let me illustrate.

Many of the early leaders in Irish agricultural development have come from farms like that of my childhood where success in living demanded independent enterprise and hard work. Barriers to be overcome in the wake of World War II were those classical conditions of West European agriculture: e.g. inefficiencies of production, technology, and fertilizers; fluctuations in produce markets; inadequate size or level of specialization to permit economies of scale. Having suffered through these constraints in their youth and having learned the values of self-sacrifice and unrelenting effort, these farmers naturally construed utopia in terms of economic rationality. The best service they could render society was to work toward a national development plan for agriculture which would enable every farmer to be free from these constraints.

The past twenty-five years has seen dramatic changes: local creameries have been rationalized, markets and fairs regionalized, transport horizons expanded and commodity prices guaranteed. Children no longer walk to school. As each individual entrepreneur and his family become more emancipated from former constraints, however, they are also deprived of former opportunities to contribute toward the collective sense of place. A bag of mixed blessings – the now familiar tension between rationality and reason – has fallen on each aspect of local life. Many residents rejoice in the disappearance of drudgery and poverty, but others wonder why one no longer sees many 'local characters' or hears any famous story-tellers, except on TV.

To translate the story into the language of home and reach, one could say that the horizons of technological and economic reach have expanded so quickly and so individualistically that one can no longer find support or centeredness from the older common-held conceptions of shared reach. Even within the lives of individual families there is often dissonance between the 'reach' capacities of members and an increasing social distance between neighbors. Also, from an ecological viewpoint, there is little thought for the long-term consequences of the growing dependence on supermarket food, artificial fertilizers, and large-scale production aimed at distant markets. Plastic bags, tin containers, and empty bottles are becoming eyesores around many homes.

If one were to ask what the meaning and potential significance of place would be here, or for similar places, one needs to redefine what is meant by community and place. One has to see both of them in dynamic terms, as horizons for basic life processes rather than as artifacts or nouns. The creation of community cannot today rely on former props like sharing limited resources and confronting together some common challenges at harvest time. Today our common challenge – still implicit rather than explicit – is a different kind: the psychological and emotional consequences of fragmented *genres de vie* juxtaposed in physical space but strangers in social space. Catalysts for community creation in rural Ireland still survive: football teams, local newspapers, common occasions of worship and recreation. The ubiquitous pub still provides the bases for contact between people and nurturing the sense of local identity.

The creeping paralysis which rural Ireland shares with many similar regions in Western Europe – from a *place* point of view – is the general penetration of Eurocratic policy in agriculture which fosters not only ecological and social scandals in land speculation but also fills the countryside with a motorized caste of development interests seeking to enlist each individual family in remote-controlled nets of consumerism. By accent, humor, and song, one can still differentiate Corkonians from Kerrymen. In overt 'geographical' terms, however, there is little difference.

Personal identity and health require an ongoing process of centering – a reciprocity between dwelling and reaching – which can find its external symbolic expression in the sense of place or regional identity. Decentralization efforts from above need to take account of this fundamental feature of lived experience both for individuals and for society. There is a need for would-be consultants in planning efforts to take cognisance of their taken-for-granted assumptions about the relationships between people and place. Ultimately, blueprints of centralization can only touch the externalities of local life; the long-term efficacy of these plans depends upon the re-awakening of creative energies in local communities.

A viable style of life for local areas cannot cling to old Newtonian conceptions of community and region, or adopt a competitive 'ghetto' attitude toward society and national spatial organization. A style of community life oriented toward self-education regarding the constantly changing horizons of reach for people, activity systems, and technology, could be a powerful catalyst for developing civic habits of sharing and discovering how much the health and happiness of individuals and communities can be enhanced by allowing people to contribute to the whole.

References

Adorno, Theodor W. 1973. *The Jargon of Authenticity*, trans. by Knut Tarnowski and Frederic Will. London: Routledge & Kegan Paul.

Bachelard, G. 1958, tr. 1964. *The Poetics of Space.* Boston: Beacon Press.

Buttimer, A. 1972. 'Social Space and the Planning of Residential Areas,' *Environment and Behavior, 4,* 279-318.

Clay, G. 1974. *How to Read the American City Close Up.* New York: Doubleday.

—, 1969. 'Remembered Landscapes.' In Paul Shephard and Daniel McKinley (eds.), *The Subversive Science,* pp. 133-9. New York: Houghton-Mifflin.

Dominian, Leon. 1917. *The Frontiers of Language and Nationality in Europe.* New York: American Geographical Society.

Haig, R.M. 1927, 1929. *Major Economic Factors in Metropolitan Growth and Arrangement,* 2 vols. New York: Regional Plan of New York and its Environs. 1927.

Howard Ebenezer. (1897 orig.) 1951. *Garden Cities for Tomorrow.* London: Faber.

Jacobs, Jane. 1961. *The Death and Life of Great American Cities.* New York: Random House.

Newman, O. 1973. *Defensible Space: Crime Prevention Through Urban Design.* New York: Collier.

Norberg-Schulz, C. 1971. *Existence, Space, and Architecture.* New York: Praeger.

Relph, E. 1976. *Place and Placelessness.* London: Pion.

Rose, Courtice. 1977. 'The Notion of Reach and its Relevance to Social Geography,' PhD dissertation, Clark University, Worcester, Mass.

Schutz, A. 1962. *Collected Papers,* 2 vols. The Hague: Nijhoff.

—, 1973. *Structures of the Lifeworld* (T. Luckman ed.). Evanston, Ill.: Northwestern University Press.

Seamon, D. 1979. *A Geography of the Lifeworld.* New York: St Martin's Press.

AFTERWORD: COMMUNITY, PLACE, AND ENVIRONMENT

David Seamon

As Anne Buttimer emphasizes in her Introduction to this volume, the period 1970-7 at Clark University's Graduate School of Geography was a special time for many people. The year 1970, when Anne arrived at Clark as a post-doctoral fellow and I as a first-year graduate student, marked the inauguration of the Core Course – a one-semester class for all first- and second-year graduate students. Led that first year by Gerry Karaska, we read David Harvey's recently published *Explanation in Geography* (1969). Class discussions were intense and stimulating, as we considered such topics as the nature and assumptions of science, the role of methodology in research, the place of geography in the social and behavioral sciences.

By the end of my first semester at Clark, I understood that perhaps the aim of graduate education is not so much to find answers as to provoke questions. As students came to Clark in the years 1970-7, they could not help but find at least some of these questions provoked by Anne or the people working with her. Through her breadth of knowledge and lucid but gentle manner of expression, Anne brought many students' attention to the taken-for-granted assumptions and world views underlying the theories and methods of social science. Over time, individuals under her guidance became a group which could often be found together over lunch at the nearby luncheonette that was then called *Bove's.* How much of what appears in the preceding pages was conceived or clarified over the rickety counters of that small corner restaurant!

These sessions involved the usual share of complaining and gossiping that is a necessary and integral part of any graduate student's life. At the same time, however, there was serious discussion on all variety of themes. For instance, I remember the day that Torsten Hägerstrand came to lunch with us: there was extended argument over the relative amount of understanding gathered from a translated text versus the original. At first glance such a theme may seem a strange topic for geographers, but it indicates well the range and variety of our conversations.

Perhaps the theme most frequently resurfacing in our discussions, partly because of Anne's growing disenchantment with the perceptual approach to environmental behavior and her rising interest in existentialism and phenomenology, was the question of how the behavioral and experiential relationship between people and place could be accurately portrayed. Should behavior be emphasized, since it is empirically observable and verifiable, while experience is not? Should behaviors of specific individuals be studied in qualitative fashion or should behaviors be aggregated into units and examined mathematically? Was the perceptual claim that cognition structured behavior accurate? How would alternatives to conventional scientific approaches, particularly existentialism and phenomenology, speak of environmental behavior and experience?

Over time, each student working with Anne had to deal with these and related themes in his or her own way; the essays in this book indicate the range of solutions attempted. Though our approaches were often unquestionably different, we also – perhaps more deeply than we realized at the time – shared some common themes and points of view. This afterword indicates some of these commonalities as I've come to see them. I do not claim to speak for the other contributors, who each would no doubt interpret similarities in a considerably different fashion; nor do I claim that each of the themes discussed is present in all the articles.

The work described in this book offers approaches to geographic research which are novel, and to some skeptical readers, perhaps more sociological, psychological, or philosophical than geographic. On the other hand, one senses, particularly in today's younger students, a growing frustration with conventional approaches and theories of social and behavioral sciences, which often seem hopelessly out of touch with the events and needs of daily living. My belief is that the work done under Anne at Clark during 1970-7 indicates some valuable new directions for social science research in general and geography in particular. I hope that an attempt to elucidate some of the common elements of this work – as incipient and personal as the sketch may be – might give old and new students alike some ideas for research alternatives and possibilities.

Fostering Receptivity

An openness to the subject of study as an explicit part of the research method is one theme illustrated by some of the preceding essays. This way of working arose in the Clark group largely from the impact of phenomenology. Already in her 1972 'Social Space' article, Anne expresses a need to reach beyond the many arbitrary, piecemeal approaches to urban residential behavior and to arrive at a perspective more in tune with the lives and worlds of people residing in actual urban places. 'We need frameworks for investigation and reflection,' she writes, 'which do not segment or ossify parts of the city as Cartesian practices have done.' Rather, she seeks a picture of urban living and life-styles *as they are in themselves,* authored by real people. The aim is 'an empathetic understanding of urban life as existential reality, as lived experience.'

By 1978, in the essay 'Home, Reach, and a Sense of Place,' Anne can look back at the 'Social Space' article and recognize that it 'was not an objective interpretation.' Though it intended to establish an openness to the lives of the Glasgow housewives studied, the study was flawed, she says, because there had been an unintentional manipulation of the subjects and their responses through the filters of her own experience. The need becomes an approach to elicit and interpret these and similar experiences in a way whereby they appear as they are in themselves rather than as the researcher makes them out to be because of theories, assumptions, personal background, or some similar set of blinders.

One aim of the Clark group has been to discover channels for promoting openness. How can one find an approach to see the thing as it is in its own fashion? How can one have subjects present their world as it is for them? How can one understand that world in the subjects' terms and not his or her own?

Curt Rose discusses this need for receptivity most directly when he argues that human geography is the interpreting of texts. A text, as he defines it, relates not only to written artifacts but to all other domains of order such as material objects, landscapes, gestures, or behaviors in places. The aim is to uncover the text's meanings, both those openly expressed as well as those 'which are only implicit in the text taken as a whole.' Receptivity to the text is a primary need in this interpretive process. How can one let the text speak so that it can be interpreted in terms accurately portraying it?

In one sense, it can be said that some of the preceding essays represent attempts to interpret texts which involve the realm of environmental behavior and experience. Graham Rowles works to set aside conventional gerontological perspectives and to re-interpret elderly people's geographical experience from in-depth descriptions. Mick Godkin attempts a similar interpretation for alcoholics and the meanings which place might have for them. I explore the essential structure of everyday human movement, working to separate from the conventional cognitive and behaviorist approaches. The result in each of these studies, I think, is an innovative point of view outside the reach of conventional perspectives and theories grounded in the usual logical-positivist stance.

Presently, research in behavioral and social geography is dominated heavily by a cognitive perspective: environmental and spatial behavior is interpreted to be a function of cognitive image. These three essays indicate the importance of other experiential dimensions: for example, feelings and fantasies in relation to place; the role of the body in spatial behavior; the importance of stability, continuity, and a sense of belonging in relation to one's environment. These themes are new to behavioral and social geography and arise partially because the researcher has worked explicitly to foster an atmosphere of receptivity in his or her work.

Exploring Lifeworlds

Lifeworld is the taken-for-granted context and pattern of daily living (Buttimer, 1976). Normally, people are busy with day-to-day events and situations; there is no time nor wish to explore the lifeworld explicitly for meaning.

Much of the Clark work has attempted to clarify the nature of lifeworld, though not all the essays use the notion explicitly. Bobby Wilson's dissertation (1974) examined the experience of migration among black families in Bedford-Stuyvesant, New York. In interpreting his data, he sought a marriage between social space and the theory of symbolic interaction. His article summarizes this relationship and offers a picture of place behavior and experience founded in a theoretical framework making social fabric the independent variable.

Later work, emphasizing the need for openness, sought to avoid *a priori* theoretical structures and to let the lifeworld emerge in its own fashion. There might be interest in a specific lifeworld – the elderly's or alcoholic's – or concern in more general terms – exploring the nature of *reach* (Rose, 1977), or the nature of *everyday environmental experience* (Seamon, 1977). Though she approaches her research from an anarchist perspective and does not use the term 'lifeworld' directly, Myrna Breitbart describes day-to-day change resulting from the anarchist movement in Civil War Spain. This study is exciting because it highlights a dramatically shifting lifeworld and some of the improvements in peoples' daily living.

The examination of lifeworlds is significant to geographic research because it reveals dimensions of daily living which people normally take for granted and do not make objects of conscious attention. Notions such as rootedness, uprootedness, reflective geographic fantasy, and body-subject each articulate experiential and behavioral structures which normally are matter of fact and therefore concealed. Such notions can be the foundation for alternative perspectives and theories in behavioral and social geography. At the same time, their understanding helps the student to become more attuned to his or her own daily living and the places and environments where it unfolds.

One dimension of lifeworld having particular value for residential planning and policy is the notion of *insiders and outsiders* discussed by Anne in her second essay. The insider's world is one of process and events normally unnoticed and unquestioned. He or she *dwells* in place and rarely conceives it as an explicit entity which might be made the object of directed attention. In relation to planning and policy, the trap for the insider is that 'one lives in places and may be so immersed in the particulars of everyday life and action that he or she may see no point in questioning the taken-for-granted or in seeing home in its wider spatial or social context.'

In contrast, Anne goes on to say, the outsider views place largely in terms of artifacts – for example, land use, housing, activity flows, number of services. The assumption is that manipulation of the material environment will lead to a more livable place. The trap here, however, is that the outsider views place 'from an abstract sky. He or she tries to read the texts of landscapes and overt behavior in the picture language of maps and models and is therefore inevitably drawn toward finding in places what he or she *intends* to find in them.'

Anne believes that sensitizing both insider and outsider to the aspects of lifeworld could serve two functions: help the insider to see his place as an explicit entity with links to a larger socioeconomic milieu; help the outsider to supplement his 'noun' language with the processual dynamics of place as lifeworld. In this way, there may be better communication 'between those who wish to live in places and those who wish to plan for them.'

Emphasizing Understanding

Much of the past work in social science, attempting to imitate the logical-positivist stance of natural science, has emphasized *explanation* – the articulation of the factors causing an occurence (Grange, 1974). Explanation seeks to find out *why* something happens, generally by relating sets of variables. For example, the researcher demonstrates (usually through statistical validation) that residential satisfaction is related to socio-economic background, psychological characteristics, qualities of the physical environment, or a similar set of variables.

Some of the preceding essays promote an alternative to explanation – *understanding,* as Graham Rowles calls it. Understanding involves a looking and seeing that immerses the researcher in the theme he studies. He becomes more 'at home' with the person or thing, and sees him, her, or it with more concern and empathy. His discoveries may have meaning for his own personal life and help him to appreciate in more heartfelt fashion the world in which he lives. Understanding has been de-emphasized in most past social science research or it has arisen only as an unintentional by-product from the primary search for explanation. The researcher has studied his subjects as objects separate from his own personal world. Identifying with their situations or becoming emotionally involved with their needs has been seen as 'subjective' or 'poor research procedure.' They are weaknesses to be avoided.

In contrast, Graham's work, as well as Mick Godkin's and mine, seek to enter into the world of the subjects, who might be more accurately called 'co-researchers.' These participants gain insights into their own day-to-day life as they provide descriptive material for the researcher. The result, Graham explains, is a deeper level of awareness 'which arises from drawing close enough to a person to become a sympathetic participant within her lifeworld and to have her be integrally involved in one's own.'

In arguing for a dialogue between insider and outsider, Anne argues that the main task may be pedagogical and involves 'a calling to conscious awareness those taken-for-granted ideas and practices within one's own personal world and then to reach beyond them.' Underlying this attunement to lifeworld is the fostering of understanding in relation to the role of place and environment in one's day-to-day living. Several notions introduced in the preceding essays – for example, home, reach, time-space routine, and rootedness – could provide valuable concrete foci for directing such understanding.

Linking People, Place, and Community

Embarrassed by environmental determinism and shifting emphasis to geography as a spatial science, geographers in the last several decades have been hesitant to explore the precise behavioral relationship among people, place, and community. The general attitude has been that the geographic environment has a role in human behavior but a role less significant than social, cultural, or economic factors. As a result, geographers have generated few original theories from within the field, but, rather, have borrowed approaches from other disciplines – particularly economics, sociology, and psychology – and explored their spatial and environmental consequences.

In part, this trend in geographic research has been affected by the times. As developments in science and technology allowed people to dominate and even dispense with environment and terrestrial space, it appeared that human beings could escape from place; that they were infinitely adaptable and could mold their relationship to place and space as they saw fit. Human living, as the sociologist Webber (1970) phrased it, was now 'community without propinquity.'

In the last decade, however, there are growing indications that the so-called 'conquest' of geographical space and environment may not be permanent. Trends of alienation, homelessness, and callous disregard for nature, coupled with the crises in energy and ecology, suggest that people are not as readily able to dispose of place and environment as they thought before. The place-environment component of the lifeworld may be equal in value to the social, economic, and psychological dimensions that have received more academic attention in the last several decades. In short, one can again say without embarrassment that people are as much *geographical* beings as they are social, cultural, or economic.

In a modern form, geographic research on the person-place bond involves at least two foci of attention: first, the *ecological* dimensions of the bond; second, the *behavioral and experiential* dimensions. An ecological focus asks how people-in-places work as ecological units. A key question is whether rootedness in place promotes a more efficient use of energy, space, and environment than today's predominant place relationship which emphasizes spatial mobility and the frequent destruction of unique places (Relph, 1976). A behavioral focus, of which the preceding essays offer a glimpse, asks a complementary question: what are the existential advantages and disadvantages of place-bound lifeworlds? Do they, for example, facilitate in better measure than a physically dispersed lifeworld such qualities as at-homeness, sense of place, care and concern for environment, community participation?

In Anne's terms of home and reach, the overriding question for both foci is, what ecologically and existentially is a suitable balance between home and reach for particular individuals and groups? Have Western people presently extended themselves too far at the expense of home? Are they presently too footloose at the expense of community stability and continuity? How can technological devices such as transportation, cybernetics, and mass communication be used to serve home as well as reach? What technologies promote dwelling and a sense of place rather than homelessness and placelessness?

Some of the preceding essays offer preliminary insight into these questions, emphasizing in their answers the importance of a secure home base in people's environmental experience. Mick Godkin indicates the essential role of rootedness and uprootedness in the alcoholic's life. Both of these patterns, he suggests, are strongly grounded in place. Similarly, Graham Rowles points to the importance of feelings in relation to place for the elderly, as well as the value of a stable home environment founded in memories of the past. My essay argues that people are inescapably bound to place partly because they are bodily beings. Habitual physical interactions, out of which sometimes arise what I call place-ballets, may be a primary grounding for a sense of community and place. Myrna Breitbart's essay is particularly interesting in relation to the person-place bond because it describes the significance of place in the anarchist conception of living. 'Decentralism,' says Paul Goodman in the essay's opening quotation, 'does not involve geographical isolation but a particular sociological use of geography.' Perhaps the person-place bond provides one point of contact among Marxist, anarchist, and phenomenological perspectives on environment and place.

Anne suggests that the most serious long-term consequence of the demise of place involves a question: 'If one can not practice a giving and receiving in one's normal everyday routines is society not denied the precious input which only individuals *in situ* could contribute?' In one sense, people *in situ* are the essence of place. As a growing malaise spreads through the Western world – no doubt in part because people feel they can no longer 'give and receive' – and as the disruption of energy and environment grows greater, circumstances may necessarily require a human return to place. If this dramatic change in life-style should occur, geographic research founded in openness, lifeworlds and understanding might not only provide a planning and policy function, but also work to renew people's awareness of their inescapable links with geographical environment, space and place.

References

Buttimer, Anne. 1976. 'Grasping the Dynamism of Lifeworld,' *Annals of the Association of American Geographers*, Vol. 66, no 2, 277-292.

Grange, Joseph. 1974. 'Lived Experience, Human Interiority and the Liberal Arts,' *Liberal Education, 60,* 359-67.
Harvey, David. 1969. *Explanation in Geography.* New York: St Martin's Press.
Relph, Edward. 1976. *Place and Placelessness.* London: Pion.
Rose, Courtice, G. 1977. 'The Concept of Reach and its Relevance to Social Geography,' PhD dissertation, Clark University, Worcester, Mass.
Seamon, David. 1977. 'Movement, Rest and Encounter: a Phenomenology of Everyday Environmental Experience,' PhD dissertation, Clark University, Worcester, Mass. Published as *A Geography of the Lifeworld.* London: Croom Helm; New York: St Martin's Press, 1979.
Webber, M.M. 1970. 'Order and Diversity: Community without Propinquity.' In H.M. Proshansky, W.H. Ittelson, and L.G. Rivlin (eds.), *Environmental Psychology: Man in His Physical Setting,* pp. 533-49. New York: Holt, Rinehart & Winston.
Wilson, Bobby. 1974. 'The Influence of Church Participation on the Behavior in Space of Black Migrants within Bedford-Stuyvesant: a Social Space Analysis,' PhD dissertation, Clark University, Worcester, Mass.

NOTES ON CONTRIBUTORS

Myrna Margulies Breitbart, Department of Social Sciences, Hampshire College, Amherst, Massachusetts.

Anne Buttimer, Graduate School of Geography, Clark University, Worcester, Massachusetts.

Michael A. Godkin, Department of Family and Community Medicine, University of Massachusetts Medical School, Worcester, Massachusetts.

Courtice Rose, Department of Geography, Bishop's University, Lennoxville, Quebec.

Graham D. Rowles, Department of Geology and Geography, West Virginia University, Morgantown, West Virginia.

David Seamon, Department of Social and Economic Geography, Lund University, Lund, Sweden.

Bobby Wilson, Department of Urban Studies, University of Alabama in Birmingham, Alabama.

INDEX

activity patterns 24
age-cohort differences 67-8
aging 49-51; and geographical experience 58-60; impacts of – community 66, family 66, personality 64-5, physical context 65, societal context 66-7; physiological variability 67
alcoholism 74-5; therapy 79-82
anarchism, Spanish 86-7, 113-16; development before Civil War 90-3; diffusion as thought and practice 93-7; impact on spatial organization 102-3; organization of work 97-102
anarchist organizations, Spanish 93-7
ateneos 93, 103
at homeness 47, 162-4, 194-5
autonomy, cultural 86

Bachelard, G. 74, 167
Beck, L.W. 124-5
becoming 22
behavioural geography 73, 84, 188, 191-2, 193-5
behaviourism 150-1, 154, 155-6
being 22
Bell, W. 136
Blumer, H. 139
Boas, F. 123
body-ballet 157-9, 160-3
body-subject 154-7, 160-3, 195
brigada 98

caja de compensacion 110
Cassierer, E. 140
Catholic Church 92
centeredness 171-2
centralization 171-2
Chombart de Lauwe, P.H. 23
cognitive maps 25, 138, 151, 153, 155-7
collectives, Spanish 87; agricultural production 101-2; regional links 107-10; spatial organization 102-7, 112-13
collectivization, Spanish 87, 97
community: and environment 86-7, 160-4, 170-2, 174, 181-2, 193-5; authentic 143
Confederation National del Trabajo (CNT) 92, 95, 97, 106, 113, 116n1
Council of Aragon 112

decentralization, political 86
delegados 110
demand anticipation 29
dialogue 47
Durkheim, E. 135
environmental experience groups 152
environmental psychology 84, 150
epoché 148, 149-50, 154, *see also* phenomenology
ethnic domain 25
existential insidedness 161
explanation 69, 192-3

Fanelli, G. 92
Federacion Anarquista Iberica (FAI) 92
fields of care 60
First International 92

game 143-4
generalized other 143
genre de vie 87
geographical experience 56
geography, as a discipline 123-5, 131-2
Giorgi, A. 149
Glasgow 27-8, 179-81; centrally located less-planned estate 41-2; centrally located planned estate 40-1; less-planned peripheral estate 43-4; planned peripheral estate 43-4
Guardia Civile 91

habit 152-4, 155-7
Harvey, D. 188
home 59, 170-2
home area 47
horizons of reach 170-2

humanistic geography 55-6

images 25
insiders 170-2
interpretation 190

Jacobs, J. 21, 160-1

Kropotkin 92

latifundia 90
Lee, T. 138
Lewin, K. 137
life space 25
lifeworld 60, 62, 148, 150, 162, 164, 191-2
lived-space 161-2
localite 30
Lynch, K. 137

Malatesta, E. 115
Meade, G.H. 135, 139-45
membership groups 140
Merleau-Ponty, M. 155
movement 148, 150-4; behaviorist interpretation 150-1, 154, 155-6; cognitive interpretation 151-2, 153, 155-7; explored phenomenologically 155-7, 162

natural attitude 148-9, 163
neighborhood 162-4
Newman, Oscar 184

oberos conscientes 94
outsiders 170-2

patria chicas 87
phenomenological reduction 149
phenomenology 148-50, 154, 156-7, 189, 191-2, 194-5
place 73-4; and community 193-5; and education 183-6; changing approaches to 167-70; experiences of 172-4; styles of life and 174-84
place ballet 159-63, 195; definition 159
place-based community 141
place identification 47
play 153
privileged places 59

Ratzel, F. 136
Reclus, E. 92
reference groups 139-40, 141
refuges 77
reinforcement 154
Relph, E. 73
residential environments: livability 21-3; planning 21-3, 46-50
rootedness 75, 77-9

schema 58, 138
Searles, H.F. 74
selective attention 57
self 140; and social space 140-5; definition 140; development of 140-5
sense of self 81
Shevley, E. 136
Shibutani, T. 139
social area analysis 24
social reference system 26-7
social space 23, 135-6; and self 140-5; model 45, 46; objective component 136-7; subjective component 137-8
Sorokin, P.A. 23, 136
Spain, anarchism *see* anarchism, Spanish
spatial behavior *see* movement
spatial cognition 151-2, 153, 155-7
Standard Deviation Ellipse 35
Strauss, A. 137
surveillance zone 59
symbolic interaction 139-40

territoriality 25
text 124-5; and human geography 131-2; and implied events 129-31; and interpretation 125-7
time space routine 158-9, 160-1, 162
topoanalysis 167
Tuan, Y. 60, 73

understanding 68-9, 193
uprootedness 75-7
urbanite 30
urban renewal 46-50, 170

Ward, B. 162
William, M. 136
Wittgenstein, L. 132nl